# Continuous and Integral Bridges

# Other books on bridges from Taylor & Francis

**Bridge Bearings and Expansion Joints**
D.J. Lee

**Bridge Deck Behaviour**
E.C. Hambly

**Bridge Management**
**Inspection, maintenance, assessment and repair**
Edited by J.E. Harding, G.A.R. Parke and M.J. Ryall

**Concrete Bridge Designer's Manual**
E. Pennells

**Concrete Bridge Engineering: Performance and Advances**
Edited by R.J. Cope

**Creep and Shrinkage of Concrete**
Edited by Z.P. Bazant and I. Carol

**The Design Life of Structures**
Edited by G. Somerville

**Developments in Structural Engineering**
**Proceedings of the Forth Rail Bridge Centenary Conference**
Edited by B.H.V. Topping

**The Maintenance of Brick and Stone Masonry Structures**
Edited by A.M. Sowden

**Prestressed Steel Bridges**
**Theory and Design**
M.S. Troitsky

**Santiago Calatrava**
Edited by D. Sharp

**Tubular Structures V**
Edited by M.G. Coutie and G. Davies

# Continuous and Integral Bridges

Proceedings of the Henderson Colloquium
'Towards Joint-Free Bridges'
organized by the British Group
of the International Association
for Bridge and Structural Engineering

*Pembroke College, Cambridge, UK*
*20–21 July 1993*

Edited by

**BRIAN PRITCHARD**
***Consultant to W.S. Atkins Consultants Ltd***
***and Colebrand Ltd,***
***Haywards Heath, UK***

CRC Press
Taylor & Francis Group
Boca Raton London New York

CRC Press is an imprint of the
Taylor & Francis Group, an **informa** business
A TAYLOR & FRANCIS BOOK

CRC Press
Taylor & Francis Group
6000 Broken Sound Parkway NW, Suite 300
Boca Raton, FL 33487-2742

First issued in paperback 2019



ISBN-13: 978-0-419-19030-1 (hbk)
ISBN-13: 978-0-367-86406-4 (pbk)



A Catalogue record for this book is available from the British Library

Library of Congress Cataloging-in-Publication Data available

**Publisher's Note**
This book has been proudced from camera ready copy provided by the individual contributors.
The publisher has gone to great lengths to ensure the quality of this reprint but points out that some imperfections in the original may be apparent

**Visit the Taylor & Francis Web site at**
**http://www.taylorandfrancis.com**

**and the CRC Press Web site at**
**http://www.crcpress.com**

# Contents

# Contributors and participants

**Ir G. Bier**
Ingenieursbureau, Nederlandse Spoorwegen, Utrecht, The Netherlands

**Mr M.P. Burke Jr**
Burgess & Niple Ltd, Columbus, Ohio, USA

**Mr G.W. Clark**
British Rail Research, Derby, UK

**Mr J.R. Cuninghame**
Transport Research Laboratory, Crowthorne, UK

**Dr J.W. Dougill**
Institution of Structural Engineers, London, UK

**Mr R. Dumont**
Ministry of Infrastructure and Transport, Wallonia, Belgium

**Mr M. Fuchs**
Belgian Road Research Centre, Brussels, Belgium

**Mr A.A. Hajiarshed**
Department of Civil Engineering, University of Nottingham, UK

**Dr E.C. Hambly**
Edmund Hambly Ltd, Berkhamsted, UK

**Mr I.D. Johnson**
G Maunsell & Partners, Birmingham, UK

**Dr A. Kumar**
Kumar Associates, Beaconsfield, UK

**Mr A.M. Low**
Ove Arup & Partners, London, UK

**Mr J. Lund**
Allott & Lomax, Manchester, UK

**Professor F. Martinez y Cabrera**
Department of Structural Engineering, Polytechnic of Milan, Milan, Italy

**Dr C. Middleton**
Department of Engineering, Cambridge University, Cambridge, UK

**Mr R.J.W. Milne**
Hon Secretary, IABSE British Group, London, UK

**Mr J. Muller**
Jean Muller International, Montigny le Bretonneux, France

**Dr S. Nakamura**
Nippon Steel Corporation, Kobe, Japan

**Dr C. Ng**
Department of Engineering, University of Cambridge, Cambridge, UK

**Mr B.A. Nicholson**
Consulting Engineer, Bicester, UK

**Ms A.R.M. Norrish**
Department of Engineering, Cambridge University, Cambridge, UK

**Dr G.W. Owens**
Steel Construction Institute, Ascot, Berkshire, UK

**Mr J.D. Parsons**
Cass Hayward & Partners, Chepstow, UK

**Mr G.P. Paterson**
Bridges Engineering Division, Department of Transport, London, UK

**Mr A.J. Powderham**
Mott MacDonald, Croydon, UK

**Mr B.P. Pritchard**
Consultant to W.S. Atkins Consultants Ltd and Colebrand Ltd, Haywards Heath, UK

**Mr D.W. Quinion**
Chairman, IABSE British Group, Salisbury, UK

**Mr S. Shanmugam**
Bridges Engineering Division, Department of Transport, London, UK

**Dr Y. Shioi**
Trans-Tokyo Bay Highway Corporation, Japan

**Dr S.M. Springman**
Department of Engineering, Cambridge University, Cambridge, UK

**Mr H. Takano**
Engineering Department, Metropolitan Expressway Public Corporation, Tokyo, Japan

**Dr H.P.J. Taylor**
Costain Building Products Ltd, Stamford, UK

**Dr D.R. Thompson**
British Rail Research, Derby, UK

**Dr S. Thorburn**
Thorburn Ltd, Glasgow, Scotland, UK

**Dr G.P. Tilly**
Gifford & Partners, Southampton, UK

**Mr E.M. Veje**
COWIconsult, Lyngby, Denmark

# Preface

Welcome to the proceedings of the British Group of IABSE's annual colloquium which this year concerned joint-free bridges. The first of these colloquia was held in 1974 during the Chairmanship of Dr William Henderson from 1967 to 1979. In appreciation of his contribution the British National Group has named them 'Henderson Colloquia'.

The subject for a colloquium is selected as being of current concern and worthy of debate over 2½ days by some 30 contributors cloistered in Pembroke College, Cambridge. The contributors are invited on the basis of their experiences and the views they are likely to have, and come from several countries to maximize the benefits to be gained by all. The informal arrangements encourage discussion and debate and also provide introductions to new colleagues. The subjects in the past have dealt with design methods, materials, construction practices, professional liability and with types of structure, such as surface-stressed building structures in 1992. The proceedings are published to make the information and views expressed available to all interested people and are later discussed at a joint meeting organized by the British Group of IABSE and the Institution of Structural Engineers.

On behalf of the British Group of IABSE I thank all the participants for their contributions to the colloquium and the spirit of camaraderie which developed. We are all grateful for the preparation of the colloquium by Brian Pritchard and his organizing committee as well as Bob Milne who made it all happen.

D.W. Quinion
*Chairman, British Group, IABSE*

# Introduction

The late Bernard Godfrey, fondly remembered as a very active member of the (then) British Constructional Steelwork Association, besides IABSE, visited America in 1989. He returned full of enthusiasm for the examples of the new joint-free type bridges he had come across, particularly some very long viaducts in Tennessee. At the next IABSE British Group meeting he proposed that the 1993 Henderson colloquium should be on the subject of joint-free bridges. Bernard's proposal was immediately accepted and he started to recruit his committee. Sadly, Bernard died several months later.

Responding to Bernard's enthusiasm, a programme entitled 'Towards joint-free bridges' was put together by an organizing committee consisting of:

B.P. Pritchard (Consultant, W.S. Atkins Consultants Ltd) Chairman
A.M. Low (Ove Arup & Partners)
G.A. Paterson (Department of Transport)
D.W. Quinion (Chairman, British Group, IABSE)
Dr G.P. Tilly (Gifford & Partners Ltd)
R.J.W. Milne (Institution of Structural Engineers) Secretary

The first five named individuals also acted as chairmen and reporters for various sessions. Mr S. Shanmugam (Department of Transport) also helped by reporting on two of the sessions.

The nature of the colloquium is unique, in that attendance is by invitation only, with each of the 30 delegates, including representatives from eight overseas countries, contributing a paper. Only short introductions are given by authors and the main purpose of the colloquium is hopefully fulfilled by the lengthy discussion periods allocated after each session.

Bernard Godfrey's enthusiasm reflected the growing world-wide interest in joint-free bridging. Joints are a continuing problem in existing bridging not just because of their own failures and maintenance problems, but because of the enormous amount of corrosion damage, measured in billions of pounds, which can be caused to the underlying substructures by leakage through the joints of deck run-off water containing corrosive de-icing salts. It is not, therefore, surprising that the UK Department of Transport are currently favouring multi-span bridge deck continuity and, indeed, joint-free integral bridges up to some 70 m long. It is also evident that longer and longer integral bridges are being constructed in Europe and America.

With these factors in mind, the organizing committee invited representatives from all sectors of the bridging industry and particularly from countries heavily involved in integral bridge construction. After much deliberation, it was decided to divide the two-day colloquium into eight sessions, having the following themes:

1. Problems of bridge articulation
2. Advantages of bridge continuity
3. Continuous bridges
4. Integral bridges
5. Long-length continuity
6. Retrofitting
7. Precast beam deck continuity
8. Soil–structure interaction

In the event, the colloquium attracted excellent papers in all sessions, with lively and informed discussion inevitably exceeding the time allotted.

It is hoped that this book conveys these qualities, which were very evident to those who attended the colloquium. It cannot convey, however, the spirit of the event – a heady mixture of international camaraderie and expert first-hand knowledge overlain with the nostalgia for student days which Cambridge generates.

May I conclude this introduction by recording my heartfelt thanks to the participants and to the hard-working members of the organizing committee, who all contributed so much to this successful event. I must also thank Bernard, no doubt in a bridge-decorated afterlife, for pointing the way.

*Brian Pritchard*

# PART ONE

# PROBLEMS OF BRIDGE ARTICULATION

# 1 PERFORMANCE OF DECK EXPANSION JOINTS

I.D. JOHNSON
G. Maunsell & Partners, Birmingham, UK

Abstract
The infrastructure developments of the 1960's led to a considerable increase in the UK bridge stock. It is recognised that the maintenance of this stock now represents a major annual expenditure which has been increased by the combination of the use of rock salt for de-icing and the deficient performance of expansion joints.

Due to concern expressed by the Department of Transport regarding bridge deck expansion joints the Transport Research Laboratory appointed G. Maunsell & Partners to carry out a study of the condition and performance of 250 highway bridge deck joints in-service. The commission was awarded in February 1990 and completed in July 1992.

This paper is based on the findings of the study and discusses the selection, specification and detailing of joints and associated components.

Inspections of the joints indicated that a substantial proportion has some form of defect and that a significant proportion leak. Bearing shelves and sub-surface drainage systems were found to be poorly detailed or badly maintained in a large number of cases and this exacerbated the effects of the leakage of run-off water through expansion joints.
Keywords: Bridge, Expansion Joint, Performance, Selection, Specification, Detailing.

## 1 Background

Several studies have been conducted over the last decade into the durability of highway bridges, Booth et al (1987), Weisgerber et al (1987), PIARC (1987) and Wallbank (1989). All of these studies, regardless of the country involved, highlight expansion joints as being critical elements affecting the maintenance-free lives of structures and include recommendations that specific studies be conducted on the performance of joints in-service. The most recent national investigations of expansion joints in the UK prior to this study were conducted by Price (1984).

Until the mid 1970's the most common forms of expansion joint in use in the UK were the buried and nosing types, however, as a result of the failure of the surfacing over buried joints and the generally poor

*Continuous and Integral Bridges.* Edited by B.P. Pritchard. Published in 1994
by Taylor & Francis, 2 Park Square, Milton Park, Abingdon, Oxon, OX14 4RN. ISBN: 0 419 19030 9.

performance of nosing joints, the asphaltic plug joint (APJ) has now been widely adopted.

The original APJs, developed in Italy and the UK during the late 1970's, suffered from an unacceptable degree of tracking during periods of hot weather and, although there have been continuing modifications of the bitumen binders by the manufacturers, it is noticeable that the present day performance of APJs is not wholly satisfactory.

The work by Price included a limited survey of APJs. The significant annual expenditure now associated with the maintenance of the UK bridge stock, coupled with the increased use of APJs, and the identification of the performance of expansion joints as critically affecting the durability of highway bridges lead to the Transport Research Laboratory appointing G. Maunsell & Partners to conduct a study of the condition and performance of expansion joints in-service.

The study was intended to update the available knowledge concerning the performance of the more commonly used types of joint, namely APJs, reinforced elastomeric joints (REJs) and elastomeric in metal runner (EMR) joints and involved the inspection of 250 joints over a period of 18 months. Full details of the joint types studied, procedures used and the locations of the joints surveyed are included in the final report, Johnson and McAndrew (1993).

It was found that the majority of expansion joints have some form of defect and that the problems identified by previous workers still persist. This paper is based on the findings of the study and discusses the selection, specification and detailing of joints and their associated components.

## 2 Selection

### 2.1 Performance requirements

The purpose of expansion joints is to permit movement of the bridge deck whilst ensuring that the ride quality and watertightness of the joint are not unacceptably affected. Furthermore, it is important that the joint be able to accommodate freely the design movements expected of the structure, whether they be rotational, lateral or longitudinal, without either transferring potentially damaging forces into the structure or causing distress to the joint itself.

As well as these performance requirements, the process of selection of an appropriate joint should also take into account the characteristics of the site that is being considered. Although the Department of Transport (DoT) Departmental Standard and associated Advice Note provide some guidance on the selection of joints, DoT (1989a), DoT (1989b), it could be argued that the implementation of the current UK Specification for Highway Works, DoT (1992), limits the selection process and does not necessarily enable to engineer to specify the most durable or appropriate joint type for a given situation. DoT policy with respect to the assessment of durability is currently under review and it is understood that the concept of whole life costing is to be introduced, this point is discussed further on in this paper.

**2.2 Movement**

Acceptable movement ranges for the various joint types in use on UK roads are given in the Departmental Standard, however, the relative proportions of static and dynamic components can significantly affect the performance of a joint. The acceptability limits provided for these components in relation to joint performance, particularly with respect to APJs, is less clear and is open to some conjecture.

The surveys conducted as part of the study identified that the performance of APJs was often incompatible with the relatively large magnitude of dynamic, or high frequency, displacements and rotations that can be experienced on steel composite decks when these structures are subjected to heavy traffic densities, particularly those incorporating a high proportion of heavy goods vehicles (HGVs). It was noted, however, that the cracking and leakage of APJs resulting from this incompatibility was usually confined to untrafficked areas, such as the hardshoulder or central reserve, and, to a lesser extent, to the outer lane of dual 3 lane carriageways. It is considered that this distribution of defects is due to the remoulding effect that trafficking, particularly by HGVs, can have on the binder matrix of APJs.

It was reported by maintenance engineers that cracking of APJs was often initiated after the occurrence of hard frosts. The onset of this cracking is considered to be the result of relatively sudden thermal movements associated with rapid temperature changes.

Cracking, due to either trafficking or frost, is believed to occur due to limitations in the elasto-plastic behaviour of the binder material. It is, however, considered that insufficient information is currently available to enable practical guidelines to be prepared which would then permit the design engineer to assess potential joint performance from the binder properties.

Concrete decks generally tend to be stiffer than their steel counterparts, the dynamic component of the total joint movement is, therefore, not as onerous. It is, however, essential that the static, or low frequency, movements arising from the creep and shrinkage of concrete decks are fully assessed; inadequate provision for these movements has been found to result in the dislocation of elastomeric inserts in EMR joints. Similar errors could also cause distress in REJs and may lead to splitting. Movements due to creep and shrinkage appear to be significant only on relatively long span, or multi-span continuous structures. No deleterious effects in APJs have been attributed to these movements.

**2.3 Influence of site on joint type**

The performance requirements of a joint are partially related to the characteristics of the site in which it is to be placed and the site should, therefore, influence the selection procedure. Moreover, the nature of the installation itself, whether new construction or replacement work, can place significant restrictions on the suitability of some joint types.

Heavy trafficking is a primary concern and can lead to tracking of APJs and an unacceptable ride quality. If tracking is allowed to develop unabated then increased impact loads can result in the deterioration of the adjacent surfacing and the eventual potholing of joints. Because of this, it is recognised by many maintenance authorities that regular resurfacing of the upper 40mm of APJs may be required on busy routes.

The behaviour of REJs and EMR joints can also be adversely affected by trafficking. The inspections carried out indicated that bolted connections, and particularly those anchored by the placement of a mortar bed under the complete joint, are inappropriate for routes with a high flow of HGVs. The vibrations caused by such vehicle flows have been found to result in the stripping of bolt threads and, in some cases, severe deterioration of the anchorage material. It should be noted that an anchorage bedding has been observed to reach the point of disintegration within a period of three months from the time initial signs of distress were noted; this highlights the need for the regular inspection of joints to assess their integrity.

In addition to vibration related damage, mechanical anchorages are also subjected to dynamic impact loads which have been shown to be up to 1.7 times the static wheel load, Koster (1986). These impact loads no doubt play a significant role in the deterioration of anchorage beddings as noted above. Such deterioration is not as frequent in the anchorages of EMR joints, the majority of which are cast-in and have a reinforcement cage integral with that of the deck.

Cast-in anchorage details are, however, usually wholly inappropriate for repair situations where a failed joint is to be replaced. In such instances surface mounted joints are finding much favour in the UK. These are bonded to the deck via resin nosings and although one manufacturer has gained a good track record, other systems, based on the same principle, have failed during service trials due to inadequate anchorage.

Deck joints which experience mechanical anchorage failure will usually require considerable remedial works to return the joint to a serviceable state. This is equally true of segmental bolted joints where damage to a single segment results in the need for the replacement of that component; the time taken for the repair of bolted joints can be significantly increased, and thus serviceability seriously impaired, due to the ingress of salt and grit.

The acceptability of maintenance operations, whether they be regular minor works to APJs or infrequent major repairs to failed mechanical joints, must be considered with respect to the disruption that would be caused to road users, particularly on major arterial routes such as the M25. Such considerations may demand the selection of a more reliable form of joint even if this incurs an increase in the initial capital cost.

An over-riding factor in the selection of the form of joint may be the resurfacing requirements for the main carriageway. This is particularly applicable to multi-span simply supported structures such as the elevated sections of the M5 and M6 through Birmingham. There may be a case for re-articulation of the deck and the introduction of continuity when major maintenance of such structures is carried out. G. Maunsell & Partners adopted this approach for the refurbishment of the A19 Tees Viaduct and the subject is to be covered in greater detail by another author. Such refurbishment exercises can modify the performance requirements for expansion joints, eg. the transfer of braking loads between spans. These must be taken into account during the selection and specification procedure.

The study has shown that up to 70% of the APJs and approximately 65% of the REJs inspected were leaking. In contrast, leakage through EMR joints occurred in only 40% of cases and the majority of these were related to the use of preformed sheet waterproofing systems; the use of a liquid membrane system would probably alleviate many of these cases. The expenditure on the repair and rehabilitation of the DoT stock of concrete bridges has been reported by Wallbank to be in the order of £60 million per annum, a significant proportion of which has been necessitated by the effects of chloride laden run-off water leaking through expansion joints. In cases where concrete substructures cannot be provided with protection, e.g. waterproofing the bearing shelf area and deck ends, then a heavy onus should be placed upon the performance of the selected joint with respect to watertightness.

## 3 Detailing and specifications

The selection of an appropriate joint must be supported by the conscientious detailing of the joint, drainage systems and the underlying substructure in order to ensure that a durable design is achieved.

An Advice Note on Design for Durability, currently being prepared by the DoT, Holland (1993), contains guidance on the provision of abutment galleries where conventional abutments are to be used as well as other design aspects pertinent to expansion joints.

The current UK DoT Specification for Highway Works restricts the level of detail that can be incorporated in specifications for expansion joints. Provision is made for the specification of subsurface drainage and it is strongly recommended that such drainage be provided where water is likely to pond on the bridge deck adjacent to the joint.

A large majority of the problems observed with expansion joints can be attributed to the inefficient drainage of sub-surface water. Unfortunately the drainage systems currently used either have insufficient capacity or appear to be unmaintainable and become blocked over an unacceptably short time scale. Transverse drainage systems can be detailed such that they can be rodded or jetted and such detailing should be encouraged wherever possible.

If feasible the inclusion of dual drainage systems should be considered; transverse drainage can be placed immediately adjacent to the joint and combined with through-deck drainage some distance away from the joint. Consideration must, however, be given to the discharge from such drainage to ensure that other areas of the structure, e.g the internal surfaces of box decks, are not inadvertently affected by runoff water or put at potential risk.

There is a real need for further developments in the design of subsurface drainage systems and this will become increasingly necessary with the introduction of porus asphalt surfacing. The need for such drainage can however be alleviated by the use of buried joints if articulation can be arranged such that the lowest end of the deck also acts as the point of fixity and that dynamic rotational movements are limited.

Despite the restrictions on the engineer's freedom to specify joint details, it is strongly recommended that the requirements of the Departmental Standard and Specification are clearly shown on tender and contract drawings. The adoption of such practices partially resolves the problem of expansion joints being regarded as secondary components in bridge design and should prompt both the design engineer and contractor to consider the detailing and installation of joints more fully.

The previous discussion of the parameters that should be taken into account when selecting an appropriate form of joint highlight some of the deficiencies in the existing specifications. It has been stated earlier that the engineer is currently unable to apply the full selection process when either specifying the form of joint to be used or when considering the contractor's selection of a joint. This state of affairs may be partially redressed by the introduction of whole life costing by the Design for Durability Advice Note. It has, however, been suggested that there may be insufficient cost data available to apply the method as comprehensively as may be desired, Leeming (1993). The current study has certainly shown this scarcity of data to be true of maintenance costs and of estimates of maintenace-free lives for expansion joints.

## 4 Concluding remarks

It has been found that the majority of expansion joints have some form of defect and that the problems identified by previous workers still persist.

A significant proportion of expansion joints leak and thus represent a major cause of deterioration of highway bridges.

There has been a notable increase in the use of asphaltic plug joints throughout the road network in the UK. However, the response of binders to dynamic and sudden thermal movements remains questionable and research to define both the required properties and to provide further guidance to engineers is required.

The current UK DoT Specification does not allow the full selection process for joints to be implemented by the specifying engineer and thus, at present, may lead to the installation of expansion joints which do not provide the best solution for the specific site being considered. There is, therefore, the need for a revision to the specification either to enable the engineer to select appropriate joint types or to comprehensively incorporate the selection procedure in a form of performance specification. If the latter course is to be taken, definitions of exactly what constitutes a failed joint will have to be prepared; these do not exist at present.

There is an increasing amount of interest in the adoption of integral bridges as a common design concept. There is, however, debate as to whether deterioration occurs in the surfacing at the interface between the road pavement and the bridge structure and whether an expansion joint should be provided at this location. This deterioration will probably be exacerbated by the introduction of porous asphalt on to the UK road network. The requirements for such a joint will be similar to those applicable to a conventional abutment and the inadequate selection, detailing and specification of such joints will result in expenditure on remedial works over an unacceptably short maintenance cycle.

Examples of expansion joints which exhibit satisfactory in-service performance have been observed during this study and prove that close attention to detail at the design and installation stages can avoid the defects which are commonly associated with deck joints.

## 5 Acknowledgement

This paper is based on work conducted as part of the research programme for the Transport Research Laboratory, carried out by G. Maunsell & Partners, and is published by permission of the Director.

The views expressed in this paper are not necessarily either those of the Transport Research Laboratory nor those of the Department of Transport.

## 6 References

Booth, Sweetman and Woolfe Consulting Engineers and Architects (1987). **Bridge Maintenance Survey Results, Report No. 108502.** Auckland, New Zealand

Department of Transport (1989a) **Expansion Joints for use in Highway Bridge Decks. Advice note BA26/88.** HMSO, London.

Department of Transport (1989b) **Expansion Joints for use in Highway Bridge Decks. Departmental Standard BD33/88.** HMSO, London.

Department of Transport (1991) **Manual of Contract Documents for Highway Works. Vol 1 Specification for Highway Works.** HMSO, London.

Holland, D.A. (1993) Design of bridges for durability and maintenance, in **Bridge Management 2, Inspection, Maintenance and Repair** (eds Harding, J.E. Parke, G.A.R. and Ryall M.J.), Thomas Telford Ltd, University of Surrey, pp889-896.

Johnson, I.D. and McAndrew, S.P. (1993) **Research into the Condition and Performance of Bridge Deck Expansion Joints. Project Report 9 E434A/BC.** TRL, Crowthorne.

Koster, W. (1986) **The Principle of Elasticity for Expansion Joints,** Mageba SA, Bulach (submitted for 2nd world congress on joints and bearings, San Antonio).

Lemming, M. (1993) The application of life cycle costing to bridges, in **Bridge Management 2, Inspection, Maintenance and Repair** (eds Harding, J.E. Parke, G.A.R. and Ryall M.J.), Thomas Telford Ltd, University of Surrey, pp574-583.

PIARC (1987) Expansion joints, Road Bridges Technical committee report No. 11, in **XVIII World Road Congress** (eds Drabble, G.M.W. and Mallett, G.P.), J.B. Shears and Sons, Brussels, pp 33-44.

Price, A.R. (1984) **The Performance In-service of Bridge Deck Expansion Joints, TRRL Report No. LR1104,** TRL, Crowthorne.
Wallbank, E.J. (1989) **The Performance of Concrete in Bridges, a Survey of 200 Highway Bridges.** HMSO, London.
Weisgerber, F.E. Minkarah, I.A. and Malon, S.R. (1987) Observations from a field study of expansion joint seals in bridges. **Transp. Res. Rec.,** 1118, 39-42.

# 2 RESEARCH REQUIREMENTS FOR BRIDGE DECK EXPANSION JOINTS

J.R. CUNINGHAME
Transport Research Laboratory, Crowthorne, UK

Abstract
Any type of joint is a potential source of weakness. Modern concrete and steel bridges have joints to accommodate the movements of the bridge deck relative to the adjacent pavement. Many of these joints leak, allowing water and chlorides to penetrate deck ends, abutments, piers and crossheads and cause corrosion of the reinforcing steel. The performance of current joints is very variable. Some are satisfactory, but many have service lives of only a few years and are expensive to maintain.

Joint performance is determined by a complex combination of factors. Some of the main factors are; the movements at the joint resulting from thermal variations and wheel loading, the properties of the materials used, and the installation conditions and standard of workmanship.

This paper gives an overview of research to develop a performance based specification based on approval testing. Laboratory tests are needed to simulate movements and wheel loading on new joints, and material tests need to be developed for use in quality control.
Keywords: Bridge, Expansion joints, Surveys, Fatigue testing.

## 1 Introduction

Bridge substructures move very slowly, but the deck is subject to short duration traffic induced movements, as well as those due to changes in the environment such as temperature. The amount of movement depends on the span, traffic loading and form of construction. Any type of joint in a structure is a potential source of weakness and expansion joints are no exception.

Designing out some or all of the expansion joints may be appropriate for new bridges and some existing multi-span structures, but for the great majority of existing highway bridges expansion joints will have to be maintained, and replaced several times during the life of the structure. Irrespective of the type of joint used, the requirements are simply stated.

- To provide a seal between two parts of the structure which are moving relative to one another.
- To provide a smooth, quiet, durable running surface for traffic.

## 2 Joint types

Several types of expansion joint are used in the UK to cater for different movement ranges; buried (20mm max movement), asphaltic plug (40mm), nosing (40mm), elastomeric, in metal runners and cantilever comb or tooth (mechanical joints). The maximum capacity of the latter three types varies according to manufacturer and type. Requirements for expansion joints on highway bridges are given in Departmental standard BD 33/88 (Department of Transport, 1989a), with additional advice in a companion Advice Note, BA 26/88 (Department of Transport, 1989b).

In general there are few problems with elastomeric joints. The elastomeric seal may be damaged by sharp debris, but it can be replaced. However, the majority of the UK bridge stock consists of short span bridges, asphaltic joints are more common than the larger capacity types. Buried and nosing joints were widely used in the 1970's, but asphaltic plug joints are currently the most popular type. Current research at the Transport Research Laboratory (TRL) is therefore concentrated on asphaltic plug joints (but large mechanical joints may become more widely used in future to provide a single joint in an otherwise continuous structure.

## 3 Review of Department of Transport (DOT) research

Trials and surveys of several joint types were carried out in the early eighties. These included buried joints (Price, 1982), open gap (nosing) type joints (Price, 1983) and a seven year study covering all the main types then in service (Price, 1984). Since then, buried joints and open gap (nosing) joints have largely been superseded by asphaltic plug joints and these have been further developed with the introduction of new materials. Nevertheless, many of the conclusions of that study are still valid. They can be summarised as follows.

- Joint performance is determined by a complex combination of factors which vary with and within joint types.

- A principal factor affecting joint performance is the nature and magnitude of the structural movements at the joint resulting from thermal variations and traffic loading.
- Installation conditions can severely limit the life of a joint; poor workmanship was responsible for many premature failures.
- The choice of materials can significantly affect long term performance.
- Structures with relatively flexible decks can present severe conditions for certain types of joint. Joints are also affected by bearing performance.
- Weaknesses in joint design were responsible for premature failure, mainly in the wheel tracks.
- Leakage at joints often caused damage to substructures and bearings.

The emphasis was then switched to laboratory testing.

Nottingham University were commissioned to develop a laboratory test facility for movement testing of buried joints. Joint specimens were made up in a tray, 0.5m wide by 1.0m long. The tray is in two halves, connected only by the joint material. Horizontal and vertical movements can be applied separately or in combination and rotational movement can be applied in place of the vertical movement if required. Both rapid movements due to traffic and slow changes simulating thermal variations can be applied. The specimen is housed in a temperature controlled chamber.

Tests were carried out on buried joints (Brown et al, 1991) to investigate thermal movement, traffic induced horizontal, vertical and rotational movement, and the effects of test temperature and stress condition of the joint at the start of the test. It was found that typical service failures could be reproduced in the laboratory and the performance of the test specimens was comparable with observations of service performance by Price (1982). This facility has now been installed at TRL, updated and developed to test asphaltic plug joints, see Figure 1.

At the same time, Reading University were asked to assess the feasibility of a laboratory test facility capable of applying service wheel loads in the laboratory. Many possible methods of applying wheel loading were considered. Similar equipment is used in pavement test facilities, but none of the existing designs was suitable. Rotary machines of manageable size would apply significant scrubbing forces, linear designs apply the load forwards <u>and</u> backwards which may well affect expansion joint performance, and most have too slow a cycling rate. The final design was a rotary machine with straight sections over three test areas.For reasons of cost a third scale machine was envisaged. However, a parametric study by Aberdeen University showed that a scale model could not provide the basis of an approval test. The visco-elastic properties of asphaltic joints would make it impossible to scale the effects of speed, temperature and material properties (eg aggregate size), and hence to relate the test results to behaviour at full scale. Also, it would be uneconomic to manufacture scale models of mechanical joints, and to be sure that the model would behave in the same way as a full size joint. The solution to this problem is outlined in section 6 below.

## 4 Research objectives

The current DOT approval system for expansion joints for highway bridges requires a one year trial installation period on a highway bridge, ie approval testing in the field.

Trials cannot provide reliable information on the performance of a new joint quickly enough to be useful. There are two problems. Conditions vary so widely that performance on one bridge cannot be regarded as representative, and secondly it is often impossible to identify the cause of any deterioration in service. The result is that the Engineer does not know which applications the joint may be suitable for, and the joint manufacturer has little guidance as to how to modify the joint to improve performance. There is also a natural reluctance to install untried joint types on heavily trafficked structures because of the possible traffic disruption. About 10 years of experience, and repeated condition surveys would be required to determine the performance of a joint in the full range of conditions, by which time it may be obsolete.

A laboratory based approval system has several advantages. The severity of the applied movement and loading can be varied to assess a joint for use in a variety of conditions allowing the Engineer to select an appropriate joint for each application. Joint manufacturers will have a consistent, testable requirement to design to and could modify and retest new joints quickly in the event of early failure. Also the research necessary to define a set of approval tests will lead to a better understanding of joint behaviour and the conditions which joints have to withstand in service. The disadvantage is of course that the environment (weather etc) cannot be reproduced at justifiable cost in the laboratory.

There are two basic requirements which a laboratory system will have to meet. The test must reproduce the types of joint deterioration which occur in service, and it must produce results quickly to assess new products.

The aim of current and future research is to reduce expenditure on maintenance and replacement of expansion joints, and to prevent damage to bridge substructures due to faulty joints. Specifically, the objectives are as follows.

- To determine the performance of joints currently in service.
- To develop suitable tests for type approval of joints, and for routine quality assurance testing of installed joints.
- To devise a performance based specification for bridge expansion joints.
- To develop improved design guidance and promote higher standards of installation.
- To establish agreed failure criteria and provide guidance for maintenance authorities.

## 5 Asphaltic plug joints (APJ)

As noted previously asphaltic plug joints have seen greatly increased usage over recent years and are now the most common type. They are relatively cheap and quick to install, reducing traffic management costs. However, durability is a problem, with some joints failing in a matter of months. The most urgent need is therefore to understand and improve the performance of this type of joint. A typical APJ is shown in Figure 2.

Many bridges are resurfaced after about seven years and it is convenient to replace the expansion joints at the same time. It is calculated that APJ's are economic if they have a service life of at least 5 years. This gives a target life of 5-7 years.

The performance of an APJ depends on:

- The material properties of the plug.
- The installation (procedure and standard of workmanship).
- The design of the joint, eg joint geometry.
- The movements occurring in service (thermal and traffic induced.
- The wheel loading imposed by heavy goods vehicles (HGVs).

All of these may vary widely. Experience suggests that the most important factors are the actual material properties of the installed joint (as opposed to the design values) and the standard of installation and to a lesser extent the movements and wheel loads.

Failure of bridge bearings can also affect the

expansion joints. Seized bearings will alter the position of the centre of rotation of the deck under a vehicle road and cause 'arching' of the deck instead of linear thermal movement. A partially seized bearing will tend to move in a 'stick-slip' manner, and the resulting rapid movements will be more damaging to the expansion joint.

### 5.1 Definition of failure

The main faults which occur in APJs are leakage, tracking, cracking and debonding. The details of the faults found by survey and their causes are discussed by Johnson and McAndrew (1993). However, there is no agreed definition of joint failure and maintenance practices vary. One of the problems for durability testing is that failures may be progressive and interactive.

There are two views about whether leakage constitutes failure of the joint. One is that most joints leak so the designer should make provision for water to be drained away harmlessly. The other view is, that it is the function of the expansion joint to prevent water getting to the structure below the roadway, therefore a leaking joint has failed. As usual there is no simple answer. For new designs it is good practice to provide drainage and provision for cleaning the drainage channels, but for some existing bridges this is not possible. The definition of joint failure should include an assessment of the consequences of the fault. Where a leaking joint would allow water to reach parts of the structure susceptible to chloride damage, then a leaking joint should be deemed to have failed.

A certain amount of tracking can occur without affecting the performance of the joint. However, serious tracking can lead to impact loads which in turn cause deterioration in ride quality and damage to the joint and adjacent surfacing. A limit on the acceptable surface profile across the joint, eg depth of tracking, would seem to be required. It may be possible to base this limit on the standards applied to road pavements.

Cracking of the joint material and de-bonding from the adjacent surfacing are progressive failures which are likely to lead to leakage and potholing. Surface cracks may be detected by visual examination and should be repaired to prevent further deterioration. However, a joint which appears perfectly sound on the surface may leak due to subsurface faults. Water may collect in the adjacent surfacing, and traffic loading can generate sufficient hydraulic pressure to force the water past any weak point.

### 5.2 Installation

Several factors acting separately or in combination can have a very large effect on the service performance of the joint. Some of the main ones are:

- The condition of the prepared joint area. If the adjacent surfacing is wet, it can be difficult to dry the sides of the trench and achieve a good bond between the joint and surfacing.
- The cost and inconvenience of delays to traffic are becoming increasingly important and sometimes lead to great pressure on the joint installer to complete the work quickly (perhaps covering up unexpected problems instead of taking time to solve them). Re-opening the carriageway to traffic before the joint has cooled properly has also caused tracking in some joints.
- Obtaining a watertight seal between the joint and the existing deck waterproofing membrane is difficult, and may well explain why many apparently sound joints appear to leak. This problem is exacerbated by poor drainage allowing ponding and build-up of hydraulic pressure in the surfacing adjacent to the joint.

Good preparation, attention to detail and adherence to the manufacturers procedure are all necessary to achieve a durable joint. The means to achieve this are education, adequate supervision by the client, and QA testing. An example of the benefit of testing the finished product can be drawn from another area of bridge construction. It is well known in steel fabrication that a better weld is produced if the operator knows it is going to be X-rayed.

The County Surveyors Society project mentioned below will ensure that the views of client authorities are take into account. The detailed installation procedure must remain a matter for the manufacturer, but guidance can be given to promote good practice, in for example:

- Mix proportions and laying procedure - the volume and temperature of both aggregate and binder and the laying and compaction procedure.
- Environmental restrictions, eg temperature and humidity.
- The surface characteristics of the surfacing and deck, particularly dryness, roughness and the effectiveness of priming.
- The width and fixing of the cover plate and the characteristics and area of debonding.

## 6 TRL research

The Department of Transport have commissioned a programme of research at TRL to meet the objectives outlined in section 4 above. The approach to the problem will be mainly experimental, the aim is to establish relationships between:-

(a) service performance,
(b) large scale laboratory tests,

(c) type approval tests for new joints
(d) routine quality assurance tests on samples of as-installed joint material.

**6.1 Service performance**
The results of an extensive survey were reported by Johnson and McAndrew (1993) and further data on service performance will be added by a project set up by the County Surveyors Society Highway Research Group. Several other surveys have been undertaken by Local Authorities. These will help to establish the causes of failure, the maintenance practices of bridge authorities, and the costs of joint maintenance. Additional information and case studies are obtained from site visits and discussions with joint installers and maintenance authorities.

**6.2 Large scale testing**
Movement tests are in progress using the test rig developed from the Nottingham University equipment. Initial indications are that a well made and correctly installed APJ is not affected by slow 'thermal' movements at temperatures of 20°C and above. However the number of cycles of movement which can be sustained falls rapidly for temperatures below 0°C. This confirms anecdotal evidence that cracks often appear after a sharp frost.

A full scale linear wheel loading rig has been designed at TRL and is currently being manufactured. A half-axle wheel load of up to 6 tonnes will traverse a 1.5m long test section at speeds of up to 30kph, at a repetition rate of 25 passes per minute. The wheel is raised on the return pass so that the wheel runs over the joint in one direction only. The key to this machine is a spring/mass resonant driving mechanism which deals with the high accelerations at each end of the stroke, and reduces the energy requirement to manageable proportions. It is hoped to begin testing early in 1994. This rig will also be capable of testing elastomeric and large mechanical (modular) joints.

**6.3 Type approval tests**
These tests will be developed from the large scale testing. It is hoped eventually, to combine the movement and wheelloading equipment so that only a single test would be required on a new joint, with a resultant cost saving to the manufacturer.

The test load and endurance requirements could be derived from existing design wheel load spectra, for example in table 11 of the UK bridge design code, BS 5400 part 10 (BSI, 1980). European traffic loading standards are also being developed. Several spectra may be required to cover different classes of road. Given the endurance under constant amplitude loading, and a design spectrum, a fatigue assessment model can be used to relate endurance in the laboratory test to service life under traffic loading. This aspect applies to mechanical joints which may be susceptible to fatigue cracking.

Such a test (or set of tests) would be relatively expensive, and so would likely be a once only test to establish the performance of a new design of joint.

### 6.4 Material tests

A set of static and dynamic tests are being developed to establish the properties of a sample of APJ material which can be related to the properties of the specimens used in the large scale tests. Potentially suitable tests used for pavement materials are also being investigated.

A 150mm cube made up in a wooden mould is being used for these tests (rather like a concrete cube). This type of sample would be relatively easy to produce on site and so could form the basis of a quality control test. It is hoped that simple material tests could be carried out using equipment commonly available in Local Authority or commercial laboratories. The material properties likely to affect APJ behaviour are the characteristics of the binder, aggregate type and grading (which may be patented), and mix proportions.

## 7 Conclusions

Most expansion joints currently in service do not prevent the ingress of water and chlorides to parts of the bridge structure where they can cause significant damage.

Asphaltic plug joints are cheaper than other types and may perform very well in service. A service life of 5-7 years is desirable, but the performance of joints currently in service is highly variable and some fail within two years. This has led to high maintenance costs, and a programme of research is being set up to address this problem.

The objectives of the research are to identify the main factors which determine service performance and develop a performance based specification. This will be based on two levels of testing; type approval testing to supplement or replace the current trial period for the introduction of new joints, and quality assurance testing for use in joint installation.

## 8 Acknowledgements

 The views expressed in this paper are not necessarily those of the Department of Transport. The contribution of Mr R W Jordan to the research described in this paper is gratefully acknowledged.

## 9 References

Brown S F, Broderick B V and Sham K C (1991). Endurance testing of bridge deck expansion joints. TRL Contractor Report CR297.

British Standards Institution (1980). Steel, concrete and composite bridges. Part 10 code of practice for fatigue. British Standard BS 5400. British Standards Institution, London.

Department of Transport (1989a). Expansion joints for use in highway bridge decks. Departmental Standard BD 33/88. Department of Transport, London.

Department of Transport (1989b). Expansion joints for use in highway bridge decks. Departmental Advice note BA 26/88. Department of Transport, London.

Johnson I D and McAndrew S P (1993). Research into the condition and performance of bridge deck expansion joints. Transport Research Laboratory, TRL report no. PR9. Transport Research Laboratory, Crowthorne.

Price A R (1982). Trials of buried joints and surfacings on a composite motorway viaduct. Department of Transport, TRL report LR1023, Transport Research Laboratory, Crowthorne.

Price A R (1983). The performance of nosing type bridge deck expansion joints. Department of Transport, TRL report LR1071, Transport Research Laboratory, Crowthorne.

Price A R (1984). The performance in service of bridge deck expansion joints. Department of Transport, TRL report LR1104, Transport Research Laboratory, Crowthorne.

# 3 BEARING PERFORMANCE

J. LUND
Allott & Lomax, Manchester, UK

Abstract
This paper outlines the functions of bearings and reviews types of bearings. Bearing related faults are given, both those independent of the type of bearing and those associated with specific types of bearing. Access problems are given and the effects of bridge deck continuity are discussed.
Keywords: Bearing Functions, Bearing Types, Bearing Faults, Access, Continuity.

## 1 Introduction

In the UK the design life of bridges is taken to be 120 years. This does not mean that its actual life is 120 years although it is often taken to be a typical life expectancy despite many bridges having significantly longer or shorter lives. To remain serviceable throughout its life a bridge will occasionally require maintenance and repair and this includes replacement of components such as bearings, joints etc. It should not be expected that components will endure for the life of the bridge itself. Some components may be expected to last longer than others, for example bearings are expected to last longer than joints, but the impact of replacement is more severe for some components than others and generally bearing replacement may be considered to be one of the most onerous. The cost of bearing replacement may be a factor many times the cost of the bearing itself involving such expensive items as traffic diversions, temporary support to the superstructure, jacking facilities etc.

It is essential that the design of a structure not only considers the function of the bearings, but also considers the life expectancy of the bearings, access for inspection and maintenance, and ease of possible replacement. The form of structure should also be considered with regard to provision of bearings and joints. A continuous bridge deck for multi-span bridges

*Continuous and Integral Bridges.* Edited by B.P. Pritchard. Published in 1994
by Taylor & Francis, 2 Park Square, Milton Park, Abingdon, Oxon, OX14 4RN. ISBN: 0 419 19030 9.

may significantly reduce the number of bearings or remove them altogether at intermediate piers, and the so called "integral bridges" may remove the need for bearings at end supports.

## 2 Bearing function and types

### 2.1 Bearing function

The principal function of bearings is to transfer vertical and/or horizontal loads from the bridge superstructure to the substructure. However in addition they also have to accommodate rotations and translations of the superstructure relative to the substructure about either one or both of the principal axes of the bearing.

Rotations may be accommodated by :

- a) Compression of the bearing material itself e.g. rubber in either an elastomeric bearing or in a metal pot bearing.
- b) Rotation of the top section relative to the bottom section of bearing using either :
  - i) a rocking motion on an interposed metal strip or point e.g metal rocker or roller bearing.
  - ii) a sliding motion on an interposed spherical sliding plane e.g. metal spherical knuckle bearing.

Translation may be accommodated by :

- a) Sliding of the top section relative to the bottom section of the bearing by provision of a low friction interface e.g. using a PTFE/stainless steel interface (used in several types of bearing).
- b) Shear of the bearing material e.g. rubber in an elastomeric bearing.
- c) Rolling e.g. metal roller bearing.

Whilst the bearing system as a whole cannot totally restrict deck movements, they control where the movements may occur. It is the location of fixed bearings (no translation permitted), guided bearings (translation permitted in one direction), and free bearings (translation permitted in any direction) which control the articulation of the bridge, and this is a second function of bearings.

### 2.2 Bearing types

Prior to the 1950's bearings tended to be fairly basic. On shorter spans they typically consisted of either 2 plates of plain metal sliding bearings, sheets of lead or bitumen impregnated felt or just a mortar bed. On longer spans bearings were generally metal rocker or roller bearings of cast iron/ meehanite, mild or high yield steel often fabricated by a local steel fabricator. In the period following the mid to late 1950's specialist bearing manufacturers introduced new bearing types :

a) Laminated elastomeric bearings.
b) PTFE with low friction for use in the interface of sliding bearings.
c) Rubber pot and spherical bearings.
d) High strength alloys in roller bearings.
e) Fabreeka.

Modern bearings may be generalised into two main groups, elastomeric or metal, with fabreeka bearings which were very popular during the 1970's forming a third minor group, see table 1. Some bearings are a combination of elastomer and metal e.g. pot bearings, but for ease of classification these have been included with metal bearings in table 1.

## 3 Bearing faults

Bearing faults may be classified into two groups, those independent of the type of bearing and those associated with a specific type of bearing.

### 3.1 Faults independent of the type of bearing

Faults may include:

a) Design Faults
- Inadequate access or clearance preventing adequate inspection and maintenance.

b) Construction Faults
- Bearing set incorrectly or otherwise installed incorrectly.
- Irregular contact of adjoining surfaces.
- Movement of bearing restricted by some misplaced construction material or packs etc. left in place.

c) Maintenance Faults
- Leaking joints or blocked drainage pipes resulting in the bearing being continuously wet and subjected to road salts.
- Accumulation of detritus and birds nests around the bearing resulting in a damp aggressive environment.

d) Faults in the bedding mortar
- Cracking.
- Missing mortar
- Steel shims left in place creating hard spots and causing

Table 1 Bearing types

| Type | Division | Sub-Division | Comments |
|---|---|---|---|
| Elastromeric | Laminated | | |
| | Plain | Pad<br>Strip | } Restricted to small<br>} spans (<10m) |
| Metal | Roller | Single<br>Multiple | |
| | Rocker | Linear<br>Point | } Horizontal movement<br>} at free bearing<br>} provided by |
| | Knuckle | Pin<br>Leaf<br>Cylindrical<br>Spherical | } sliding interface<br>} PTFE/stainless<br>) steel<br>} |
| | Pot | | |
| | Plain sliding | | 2 plain metal plates - no longer used |
| | Fixed pin and guide bearings | | } Take horizontal<br>}loads only |
| Fabreeka | | | Horizontal movement of a free bearing provided by sliding interface of PTFE/stainless steel |
| Others | Bitumen sheet | | } |
| | Lead sheet | | } no longer used |
| | Concrete - rocker | | } |
| | Cement mortar | | } |

possible corrosion with subsequent displacement of mortar.
- Gaps between the mortar and the bearing.

e) Other faults
- Vandalism.
- Fire damage.
- Impact damage.

Apart from the access problem discussed later, the most significant fault is that of water leaking from joints above the bearings, with the problem made worse by the accumulation of detritus around the bearing holding the water. The problem of leaking joints is of course removed at piers by the use of continuous bridge decks.

Bearing locations should always be detailed to ensure good access, and bearings positioned so that they are isolated from leaking joints. They should also be detailed to enable ease of replacement in the future.

Many design, construction and maintenance faults are the result of ignorance relating to the importance of correct detailing, installation and maintenance. This could perhaps be improved by the publication of information sheets by the Department of Transport giving guidance on good practice and common defects.

### 3.2 Faults associated with specific types of bearing

Although these are not directly related to deck continuity, problems may be exacerbated by leaking deck joints and faults are briefly described in 3.2.1 to 3.2.4.

#### 3.2.1 Elastomeric bearings

Faults may include:

a) Cracking of the elastomer.
b) Irregular or excessive bulging or splitting under vertical loads.
c) Shear failure under horizontal loads.
d) Slippage of the bearing on its bed.
e) Irregular contact with the two bearing surfaces.

Elastomeric bearings are not repairable and if badly damaged they must be replaced. The elastomeric bearing itself is therefore maintenance free until replacement is necessary. Elastomeric bearings have been in service for less than 40 years and laminated elastomeric bearings are on the whole performing well, though not so plain elastomeric bearings whose use is now limited to small spans. The life of elastomeric bearings cannot be ascertained though they may well last upto 100 years or even longer.

### 3.2.2 Metal bearings

Faults may include:

a) Breakdown of the paintwork protective system.
b) Deformation of metal plates.
c) Wear on moving surfaces.
d) Indentation of plates caused by the tracking of roller bearings.
e) Cracking.
f) Over-travel of moving parts.
g) Faults relating to fixings e.g. missing bolts.
h) Transit bolts not removed inhibiting movement.
i) Sliding interface problems - see 3.2.4.

The most common fault is the breakdown of the paintwork protective system. This is made worse by the problem of leaking joints and detritus around the bearing, together with access problems preventing maintenance. Whilst metal bearings may be considered to have a life expectancy approaching the life of the bridge they require regular maintenance which if not provided (as is often the case) can lead to significant deterioration and a much reduced life.

The durability of metal bearings could be vastly improved by the use of non corrosive metals. Bearings made out of stainless steels would cost approximately four to five times those of normal steels to manufacture, but installation should cost the same and maintenance should cost less. However the number of bearings needing replacement at a cost considerably greater than the cost of the bearings themselves would be significantly reduced.

### 3.2.3 Fabreeka bearings

Faults may include:

a) Breakdown of the paintwork protective system to metal plates.
b) Slippage of the bearing on its bed.
c) Irregular contact with the two bearing surfaces.
d) Sliding interface problems - see 3.2.4.

### 3.2.4 PTFE/stainless steel sliding interface in bearings

Faults may include:

a) Rusting of the metal backing plate to the stainless steel plate so warping the stainless steel plate.
b) Dirt, scratching, minor rusting or paint on the stainless steel plate.
c) Wear of the PTFE.
d) Over-travel of the sliding interface.

## 4 Access problems

The inaccessibility of bearings creates a major problem for their inspection and maintenance. Their accessibility may be considered in two ways:

a) The accessibility of the bearing shelf: Bearing shelves are inherently inaccessible with most bearings tending to be in such locations that access plant and possibly traffic control are required to get to them.
b) The accessibility of the bearings on the bearing shelf: Often in the past little consideration has been paid to bearing accessibility and in many cases they have been positively hidden from view, presumably for appearance considerations. In even more cases, usually on concrete bridge decks supported by thin elastomeric bearings, a minimum gap has been provided between the bearing shelf and the deck soffit so that the bearings are barely visible let alone accessible.

A suitable gap should always be provided between the top of the bearing shelf and the deck soffit, and as many sides of the bearing should be accessible as possible. The size of gap is dependent on such details as the location of the bearings and the form of deck construction but may typically be in the order of 250mm.

If a bearing cannot be properly inspected it is very unlikely that it will be adequately maintained, at least until such point that its deterioration manifests itself in a problem elsewhere on the bridge. Improved accessibility may be provided at end supports by providing an inspection gallery typically 1800 mm high by 800 mm wide to the rear of the bearings, with access to the gallery often gained by a manhole to the rear or to the side of the end support. The inspection gallery has other benefits, assisting with the inspection of bridge deck joints, isolating the bearing shelf from leaking joints and providing access for other maintenance work. In spite of the benefits presented by inspection galleries there is still opposition to their use on any bridges by some authorities, whilst others are selective in their use providing them only in certain locations. On bankseats an inspection pit may be additionally provided to the front of the bearing shelf improving access to the front of the bearings.

On continuous bridge decks there is normally only one line of bearings at each intermediate support and this automatically improves access to all sides of the bearings.

## 5 The effect of continuity on bearings

The provision of continuous bridge decks over intermediate supports has several consequences on bearings given in 5.1 to 5.5.

### 5.1 Cost

The number of bearings may be significantly reduced at piers, albeit with a consequent increase in size of the bearings, resulting in a saving in both construction and maintenance costs. Typically two bearings each supporting a simply supported deck to each side of the pier may be replaced by a single larger bearing supporting a continuous deck, but this is an over simplification and exceptions may be as follows:

a) Multiple simply supported decks are likely to consist of beams, often at close centres, whereas continuous bridges may be of different form which is likely to require fewer but larger bearings across the width of the deck.
b) Piers may be constructed monolithically with the bridge deck negating the need for bearings altogether at the piers. In this case the piers are subjected to the longitudinal movements and the rotations of the deck and are limited to situations where the deck movements are relatively small or the piers tall enough and flexible enough to accommodate the movements. Even in situations where deck movements are not necessarily large, if the bridge headroom is not significantly greater than the minimum requirement then the piers may have to be so thin in order to be sufficiently flexible that they would detract from the appearance of the bridge.
c) Where deck continuity is provided by precast prestressed concrete beams landed on permanent bearings at the piers and then made continuous by a narrow insitu concrete crosshead built into the ends of the beams, the number of bearings is not reduced. However the practice of erecting beams on the permanent bearings is not common for this form of construction in the UK where the beams have tended to be landed on temporary bearings which have subsequently been replaced by a single permanent bearing located beneath the insitu concrete crosshead.

### 5.2 Effect on joints

There are no transverse joints between spans on continuous decks and consequently no leaking deck joints causing a serious problem to the condition of the bearings beneath. This is particularly important on bearings containing metallic elements susceptible to corrosion due to leaking water containing soluble road salts.

**5.3 Accessibility**
A single line of bearings on a pier supporting a continuous bridge deck gives more space for access for inspection and maintenance than two lines of bearings supporting separate bridge decks.

**5.4 Pier width**
The width of pier carrying only a single line of bearings may be reduced slightly from the width supporting two lines of bearings. This may be important for example if it is necessary to locate a pier within a standard 4m wide central reserve having standard set back and clearance to safety fences in front of the pier to UK standards.

**5.5 Bearing loads and movements**
Bearings supporting a continuous bridge deck may be subjected to totally different loadings, longitudinal movements and rotations than bearings supporting separate simply supported decks on an otherwise similar multi-span bridge. The vertical load on bearings at a pier supporting a continuous deck will be significantly greater than those of bearings supporting separate simply supported decks whilst the rotations will be significantly less. The longitudinal movement subjected to bearings will also be greater on a continuous bridge than on a bridge of separate simply supported decks unless on the latter the decks are tied. Bearings at end supports on continuous bridges are likely to be subjected to smaller reactions, smaller rotations and larger longitudinal movements than on bridges of separate simply supported decks. The consequence of this is that whereas elastomeric bearings may be viable on a bridge of separate simply supported decks they may have to be replaced by metal bearings on a continuous bridge with greater construction and maintenance costs offsetting some savings given in 5.1 above. Also bearing faults preventing the bearing accommodating deck movements may become more significant on continuous bridges with long expansion/contraction lengths. On bridges with separate simply supported decks malfunction of a bearing such that it does not accommodate deck movements does not in general appear to be resulting in structural distress of the bridge. Small deck movements associated with separate simply supported decks may be accommodated in some other manner when not taken by the bearing, for example movement of ground underlying the substructure, flexure within the substructure etc. Continuous bridges however may not be quite so forgiving of bearing malfunction, so increasing the importance of ensuring the bearing performs as it was intended.

## 6 Summary

a) There appears to be a level of ignorance appertaining to the performance of bearings relating to design, construction and maintenance. One approach to help overcome this may be the production of information sheets by the Department of Transport giving guidance on good practice and common defects. (This would not be limited solely to bearings).
b) Laminated elastomeric bearings are maintenance free and are proving to be relatively durable, but their life expectancy is not yet established.
c) Metal bearings greatest fault is corrosion, and consideration should be given to the use of non corrosive metals, although this would increase the initial bearing cost by something in the order of four to five times.
d) Bearings are often poorly maintained as a result of inadequate access and this must be addressed in the design of the bridge. An adequate gap must be left between the top of bearing shelf and deck soffit, and around the sides of the bearings. Inspection galleries should be provided at end supports to provide access for inspection and maintenance of the bearings (as well as other items) and to isolate the bearings from leaking deck joints.
e) Bearings should be detailed with ease of replacement in mind.
f) Continuous bridge decks present advantages relating to bearings in that the probable reduction in the number of bearings will reduce construction and maintenance costs, and eliminate transverse joints at intermediate supports where water containing soluble road salts may leak onto the bearings. A single line of bearings is more accessible at intermediate supports and the thickness of intermediate supports may be reduced. Continuous bridge decks are likely to result in an increase in vertical loads and horizontal movements but a reduction in rotations compared to those given by separate simply supported decks on an otherwise similar bridge. This will probably lead to larger bearings and possibly metal bearings rather than laminated elastomeric bearings.
g) Bearings with faults inhibiting translation to accommodate horizontal deck movements do not in general appear to be resulting in any manifestation of distress elsewhere in the structure where the expansion/contraction length of the bridge deck is small, thus indicating the movement is being accommodated in some other manner. This may not be the case on the longer expansion/contraction lengths found on continuous bridges, so increasing the importance of ensuring good bearing performance on such bridges.

## 7 References

British Standard 5400 : Part 9 : 1983 - Bridge Bearings

Department of Transport Standard BD20/92-Bridge Bearings. Use of BS5400 : Part 9 : 1983.

Transport and Road Research Laboratory Contract Report 260 Bridge bearing survey and problem assessment.

# 4 EFFECTS OF TEMPERATURE, SHRINKAGE AND CREEP ON INTEGRAL BRIDGES

B.A. NICHOLSON
Consulting Engineer, Bicester, UK

Abstract
This paper gives an overview of temperature, shrinkage and creep effects on bridges. Designers have to deal with these effects in different ways for simply supported spans, continuous bridges, and integral bridges. This is one of the main differences between the design of integral bridges and other bridges. Temperature, shrinkage and creep all give rise to effects which can only be calculated with a considerable degree of uncertainty, and consequently it is suggested that complicated calculations are normally unjustified. Simple methods for dealing with these effects in the design of integral bridges are suggested, and illustrated by examples.
Keywords: Integral bridges, Temperature effects, Shrinkage, Creep.

## 1 Introduction

In general, the primary loads that need to be considered in the design of any bridge are the dead loads, live loads, and wind loads. Whether the bridge has simply supported spans, is of continuous construction, or is an integral bridge, the effects of these loading groups are very similar. The design of bridges with these different forms of articulation differs mainly in the treatment of temperature and creep effects.

In this paper, temperature, shrinkage and creep effects are described in turn, and in each case the consequences for the design of integral bridges are discussed.

Two brief examples are given to illustrate some of the points raised.

## 2 Temperature effects

### 2.1 Temperature loading

BS 5400 divides temperature effects into two categories: changes in the effective temperature of the bridge, and

*Continuous and Integral Bridges*. Edited by B.P. Pritchard. Published in 1994 by Taylor & Francis, 2 Park Square, Milton Park, Abingdon, Oxon, OX14 4RN. ISBN: 0 419 19030 9.

differences in temperature through the thickness of the deck.

**2.2 Temperature differences**

Temperature difference loading causes a pattern of internal stresses to form within the deck. Hambly (1991) gives a good method for calculating these effects, and they will not be discussed further here.

**2.3 Effective bridge temperature**

Changes in effective bridge temperature cause the deck to expand and contract. As this is the most important effect governing the design of bridge joints, and also very significant in the design of bearings, it is worthwhile to summarise the way it affects bridges with different types of articulation:

A simply supported bridge span is allowed to expand and contract with little restraint. It therefore needs an expansion joint at one end, and sliding bearings to allow movement over that support. Temperature loading does not create any longitudinal forces except due to bearing restraint or friction.

A continuous bridge is also allowed to expand and contract with little restraint. Assuming it to be fixed at one abutment, it needs a large expansion joint and sliding bearings at the other end, and sliding bearings over intermediate supports. Again, temperature loading does not create large longitudinal forces.

In an integral bridge, expansion and contraction is partially restrained by the abutments, which move with the ends of the bridge. The size of the longitudinal force induced in the deck depends on the design of the abutments, and their interaction with the soil. No expansion joints or sliding bearings are required. However, there is relative movement between the abutments and the approach road, so a joint in the pavement is needed.

**2.4 Integral bridge design**

Expansion and contraction of an integral bridge will cause the abutment to move. This movement will be resisted by earth pressure behind the abutments, and possibly by friction underneath. Abutments should normally be designed to minimise such resistance to movement, as these horizontal forces have to be carried as compression or tension by the deck. Even small abutments are likely to be sufficient to resist the horizontal traffic loads.

Granular backfill should be used behind the abutments. The backfill always remains in contact with the abutments even at the extreme limits of thermal contraction, and can generate enough resistance to horizontal traffic loads with only small displacements.

There will be thermal movement between the abutments and the approach road, but integral bridges are likely to be most useful in circumstances where the pavement does not have to be interrupted by mechanical expansion joints even here. It is suggested that asphaltic plug joints should be used in the pavement at the back of the abutment, or at the end of the run-on slabs if these are used. Plug joints work well for ranges of movement up to about 25mm. Simple calculations based on the temperatures specified in BS5400 show that an all steel bridge 60m long, or an all concrete bridge 85m long, would have an overall range of thermal expansion and contraction of no more than 50mm in most locations in Britain. These bridges would have 25mm movements at each end. The vast majority of bridges have spans no more than these, so integral construction with plug joints is widely applicable. The plug joints will need maintenance from time to time during the life of the bridge. Based on experience in this country and in the USA (see NCHRP 322), this is not considered to be a problem.

Integral bridges should be designed for the same temperature ranges as others bridges. TRRL 696 and TRRL 765 show that the annual temperature range is only about 30% less than the 120 year extreme, so there is little benefit in designing for a reduced return period, particularly as it is normally straightforward to cope with the 120 year thermal movements as described above.

## 3 Shrinkage effects

### 3.1 Shrinkage

Concrete shrinks slightly as it ages, and this can affect stresses and deformations in bridges.

In composite bridges, the separate components will in general shrink by different amounts, inducing a set of internal stresses and curvature of the deck, and also a small overall shortening. The internal stresses must be allowed for in the design, and calculations take a similar form to those for temperature difference loading.

The overall shortening will be most significant in an all concrete bridge. A conservative value for long term shrinkage strain in concrete is about $200 \times 10^{-6}$, and about half of this occurs before the concrete has reached an age of 100 days. Even for an in-situ integral bridge, it is unlikely that integral behaviour of the abutments will start, or the pavement joint be made, before this age has been reached. Thus a realistic upper bound for overall shrinkage is $100 \times 10^{-6}$, or 10mm for a 100m bridge.

### 3.2 Effect on integral bridges

The overall shortening of a concrete bridge is much smaller than the range of thermal movement, and the gradual rate of movement will be swamped by the cyclical thermal movements. Shrinkage is not significant in the design of the abutments.

## 4 Creep effects

### 4.1 Creep

Like shrinkage, creep is a non-elastic deformation of concrete occurring over time. In the long term, creep increases the elastic deformation by a factor of about three. It is described by a creep factor, given the symbol $\phi$, and the creep strain is defined as $\phi$ times the elastic strain. The creep factor is time dependent, and normally has a long term value of about two, but can only be calculated with a considerable degree of uncertainty.

Creep affects the deformations of all bridges where concrete is used, but its effect on the stresses depends on the type of construction.

In a composite bridge, any stresses locked in by differential shrinkage between beams and the deck will be relieved by creep. This can be dealt with in design simply by applying a reduction factor to the shrinkage strain. BS5400 quotes a reduction factor of 0.43, which assumes that $\phi$=2.0.

Creep has a significant effect in prestressed concrete. In the case of post-tensioned construction and pre-tensioned beams, the effect is simply to reduce slightly tensioning force and the forces and stresses resulting from it. This is taken into account in the design simply by including creep strain with the other prestress losses.

The situation where design for creep is not so straight-forward is where prestressed concrete is used, and the stress pattern has discontinuities. This occurs when precast beams are used with in-situ diaphragms between spans. This situation requires special treatment.

Creep can have beneficial effects, in particular in all concrete integral bridges for which the range of thermal expansion can be reduced by creep. This effect will normally be small, as in the example in section 5.2. However, in cases where the abutments offer a high degree of restraint, forces on the abutment may be reduced by up to 30%, but creep calculations must be carried out if the designer wants to make use of this reduction.

### 4.2 Design using prestressed beams

This section relates to the design of both continuous and integral bridges, as creep affects these in the same way.

The difficulty arises because the prestress is not continuous over the supports. Shortening of the bottom flange of the beams due to creep causes the individual beams to hog, and opens up cracks at the bottom of the beams over the supports. The beams can be restrained so that these cracks do not occur, but this involves designing for a substantial sagging moment at the beam ends.

It is interesting to compare the bending moment diagrams when hogging of the beams is restrained (fully continuous construction) and when there is no restraint (simply-supported). Figure 1 shows on the left bending moment diagrams for a simply-supported span, made up from the combination of dead and live loading. On the right are the equivalent diagrams for a precast beam in a continuous structure. Dead loads are carried on the beam acting as simply-supported, continuity under live loads leads to hogging moments at the ends. Creep of the concrete leads to a uniform sagging moment, including over the supports. Even though this analysis is simplistic, it is clear that the total bending moment diagram is very similar to that for the simply-supported span. The creep restraint moment nearly cancels out any benefit from continuity.

Creep is very difficult to quantify with any degree of certainty, even in the simplest of designs. It has also been shown that there is little benefit to be obtained from designing the bridge as continuous, when an estimate would have to be made of the restraint moment due to creep. It is therefore recommended that the most practical way of designing a continuous or integral bridge using precast beams is to design each span as if it was simply-supported. Soon after construction has finished, when creep has only just begun, the restraint moment will not yet have built up, so hogging moments will occur over the supports. Reinforcement can be provided in the slab to resist these moments, calculated ignoring the effects of creep.

Bottom flange reinforcement in the beams to allow for the sagging moments over the supports is difficult to incorporate into precast beams, and has been shown above to be of negligible benefit to the design of the beam. Its only purpose is to control cracking. It has been found that cracking at the diaphragms over piers is not a serviceability problem: the slab is not affected, so there is no penetration of de-icing salts, and the cracks are not considered to be unsightly. There is therefore no reason to provide heavy bottom flange reinforcement, and no reason to attempt to complex calculations to design it. It is recommended that some nominal bottom flange reinforcement is provided, but that calculations are not normally necessary. Some cracking should be expected.

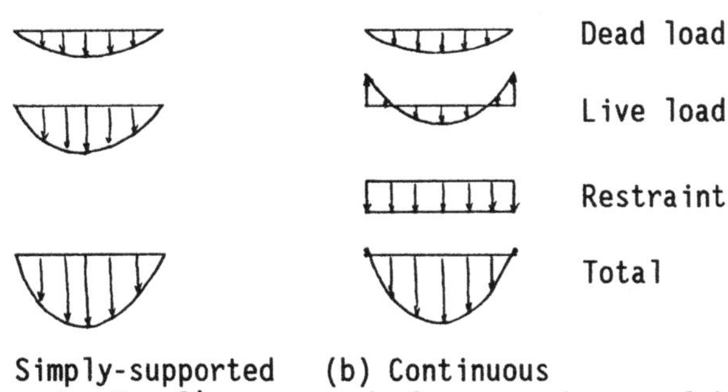

Figure 1: Bending moment for prestressed beams

## 5 Examples

### 5.1 Single span portal

Figure 2 shows a schematic diagram of a single span integral bridge with full height abutments. The deck is of composite construction, with steel beams and a concrete slab, and spans 30m.

The total range of thermal movement is less than 30mm, or 15mm at each abutment.

Plug joints are provided in the pavement at the back of the abutment.

Creep and shrinkage effects are limited to restrained shrinkage of the slab.

There is no horizontal movement between the abutments and deck, and there need also be no rotation. Bearings under the ends of the beams therefore serve no useful purpose, but would still require maintenance. It would be advantageous to omit them, and simply build ends of the steel beams into the concrete abutments. This would also apply to any intermediate supports in longer bridges, the whole structure being monolithic.

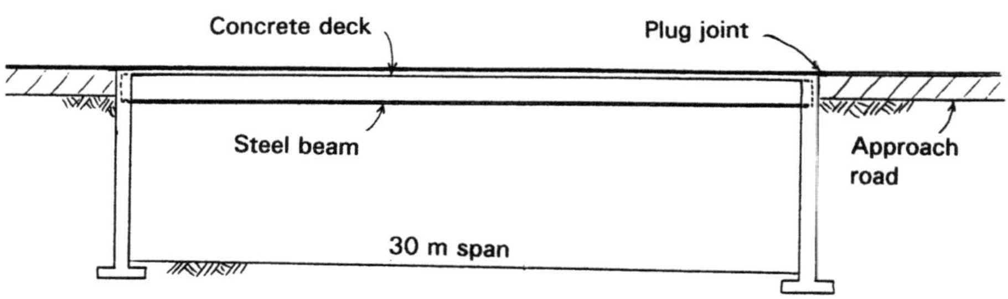


Figure 2: Single span composite integral bridge

### 5.2 Four span bridge

Figure 3 shows a schematic diagram of an integral bridge with bank seat abutments, and four spans of 15m, 20m, 20m, and 15m. The deck is of composite construction, with precast prestressed concrete beams and a concrete slab.

This example is based on Hambly and Nicholson (1990), which includes a complete set of design calculations, and also detailed calculations of creep effects.

Plug joints are provided in the pavement at the ends of the run-on slabs.

Each individual span has been designed as if it were simply-supported, and Y3 beams have been used. Nominal bottom flange reinforcement of six 20mm diameter epoxy coated bars has been provided.

Beams are placed on top of the piers with a 200mm gap. No temporary supports are required. A 400mm thick in-situ diaphragm is cast around the beam ends.

Creep will cause the beams to hog, and cracking to occur over the piers. There will clearly be some rotation of the ends of the beams, so elastomeric pad bearings have been provided.

For a total length of 70m, the total range of thermal movement is about 40mm, or 20mm at each abutment. At the extreme high temperature (24° above the mean) the thermal strain is about 290$\mu\epsilon$, which would lead to an unrestrained movement of 10mm from the mean position at each abutment. An abutment movement of 10mm would mobilise a force of about 3MN in the case of the small bank seat abutments used in this bridge. This 3MN force would give a uniform compressive stress of 0.4 MPa in the bridge, leading to an elastic strain of 13$\mu\epsilon$, which would approximately double due to creep to about 26$\mu\epsilon$. Thus it can be seen that creep might reduce the temperature expansion effects by about 10%.

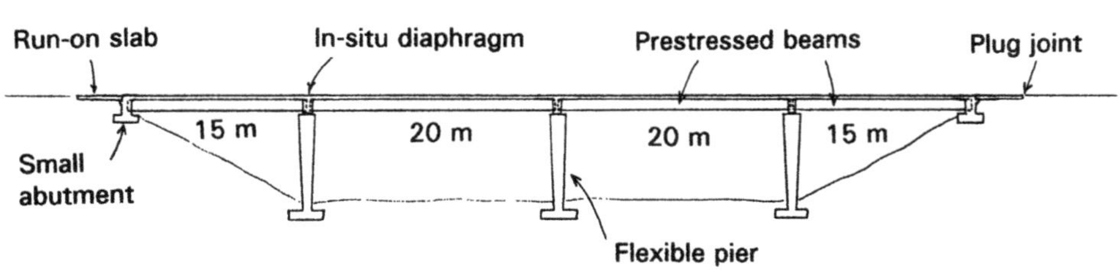


Figure 3: Four span integral bridge designed using prestressed beams

## 6 Conclusions

The main points made in this paper are:

a) The main difference between the design processes for an integral bridge and any other bridge is the treatment of thermal movements. For integral bridges the abutments have to be designed to move.
b) Integral bridges should be designed for the same temperature ranges as other bridges.
c) Asphaltic plug joints in the pavement are most appropriate for absorbing the thermal movement between abutments and the approach road for the majority of integral bridges.
d) No special treatment of shrinkage effects is required for integral bridges.
e) Creep has a significant effect on the design of continuous and integral bridges using precast prestressed beams. Any calculation of creep effects has a large degree of uncertainty, but it is suggested that no calculations need be attempted. Simple design rules have been proposed which do not require calculation of creep effects, and the penalty for making this simplification has been shown to be negligible.
f) Creep can reduce the effect of thermal expansion, but the benefit is only marginal unless the abutment offer a high degree of restraint.
g) As well as expansion joints, bearings are often unnecessary in integral bridges, and can also be eliminated.

## 7 References

BS 5400 **Steel, concrete and composite bridges**, British Standards Institution, London.

Emerson, M. (1976) **Bridge temperatures estimated from the shade temperature**, TRRL 696, Transport and Road Research Laboratory, Crowthorne.

Emerson, M. (1977) **Temperature differences in bridges: basis of design requirements**, TRRL 765, Transport and Road Research Laboratory, Crowthorne.

Hambly, E.C. and Nicholson, B.A. (1990) **Prestressed beam integral bridges**, Prestressed concrete Association, Leicester.

Hambly, E.C. and Nicholson, B.A. (1990) Prestressed beam integral bridges, **The Structural Engineer**, 68 no. 23. NB: This is a shortened version of the above leaflet.

Hambly, E.C. (1991) **Bridge deck behaviour**, E. & F.N. Spon, London.

Oesterle, R.G. Glikin, J.D. and Larson S.C. (1989) **Design of precast prestressed bridge girders made continuous**, NCHRP 322, Transport Research Board, Washington, D.C.

# REPORT ON SESSION ONE

A.M. LOW

In his presentation Mr Nicholson had proposed that in multi-span integral bridges a crack should be allowed to develop in the deck soffit over each internal support. There was concern that this crack would be over a rubber bearing where it might rip the rubber skin or it could be just away from the bearing where it might destroy the shear strength of the section. Mr Nicholson suggested that double bearings could be used, presumably with the crack between them, but he was satisfied with the single bearing proposal. It was suggested that shrinkage would tend to cancel the creep strains causing the crack. Mr Nicholson's analysis had included this compensating effect.

Mr Nicholson was asked how long an integral bridge could be and he replied that it could be any length provided there was a sufficient allowance for movement between the end of the deck and the pavement.

The question of the definition of the failure of a road joint was raised. It was suggested that the target for a joint was that it should require no lane closures for inspection or maintenance between the resurfacing operations for the road every 15 years.

The integral bridge proposal was seen as a shifting of the joint problem from the bridge to the adjacent pavement. It was pointed out that this was probably very cost effective because studies of actual maintenance costs had shown that maintenance activities for pavements could be carried out much more cheaply than similar activities for bridges.

# PART TWO

# ADVANTAGES OF BRIDGE CONTINUITY

# 5 HISTORICAL REVIEW OF THE DEVELOPMENT OF CONTINUITY AND EXPANSION JOINTS IN BRIDGES

G.P. TILLY
Gifford & Partners Ltd, Southampton, UK

Abstract
This paper is concerned with the development of bridge continuity and methods of coping with the expansive effects of temperature. Early types of bridge without expansion joints were able to cope with temperature effects by the greater flexibility of the materials and the structural forms then used. Modern construction, having high strength materials, requires articulation but it has been found that the cost of repairing failed expansion joints and the consequental damage through corrosion is a major shortfall. However, recent developments and the performances to date of long span jointless bridges has indicated that temperature problems may have been overstated and there is scope for reconsidering methods of design.
Keywords: Bridges, Temperature Effects, Expansion Joints, Continuity.

## 1 Early structures 1600BC - AD1750

The oldest surviving structure in the UK is Stonehenge, constructed in its present form circa 1600 BC using bronze age technology (Fig.1.). It was composed of 30 equal spans of 3.14m having vertical sarsen stones or piers each weighing about 30 tonne and horizontal lintels of 7 tonne. The piers had top surfaces that were slightly dished to provide a secure seating for the lintels and projecting tenons to key into mortice holes on the undersides of the lintels. In addition the lintels were fitted to each other with tongue and groove joints. Stonehenge must have represented the state-of-the-art in its time but there is no evidence of any other jointed structures of this age in the UK. It would, however, be surprising if no others were ever built and if there were no bridges.

Clapper bridges are a more primitive form of structure which may date from the bronze age albeit it has been suggested that they may be much more recent. One of the best known. the Postbridge across the Dart in Devon, is composed of three 4.6m simply supported spans and is 2m wide.

It is unlikely that the engineers of the time were aware of the pros and cons of continuity but in any case their structures were fit for the intended purpose

*Continuous and Integral Bridges.* Edited by B.P. Pritchard. Published in 1994
by Taylor & Francis, 2 Park Square, Milton Park, Abingdon, Oxon, OX14 4RN. ISBN: 0 419 19030 9.

Fig.1. Stonehenge, the oldest surviving lintelled structure in the UK.

and the few survivors have stood the test of time.

The Romans introduced arch bridges to the UK albeit the only remnants are some substantial foundations discovered at a number of river crossings. Roman arches were mainly of semi-circular shape having rises of no more than one third span. The advantage of this shape is that all the stones in the voussoir are identical and can therefore be produced very accurately. Some Roman arches were built without any mortar, an interesting example being Alcantara bridge across the Tagus in Spain. The two central arches have spans of 30m and the highest points of the bridge are 52m above the river. In 1812 during the Peninsular War one of the arches was destroyed. It was subsequently repaired but the restorers were unable to manage without using mortar.

Some of the Roman viaducts were very long indeed, for example the aquaduct at Segovia is 728m long. It is doubtful whether any consideration was given to the effects of temperature and if it had, the engineers would be unlikely to have had enough confidence to proceed.

The Roman bridge builders were supported by the advice of experts sent out by the emperors and formed into semi-military guilds. They were permitted certain tax allowances which were set against the requirement to provide advice and follow the military to distant parts of the empire. The guilds formed the basis of professional colleges in which professional standards were laid down, Shirley Smith (1954).

When the Romans departed from Britain the state-of-the-art fell back and the pointed arch was adopted. This required less skill and exerted a lower thrust on the abutments. Pointed arched were commonly used up to the fifteenth century.

During the canal era many arch structures were constructed which are still in operation today without having had major strengthening. These structures were built with lime mortar which permitted relative movements between the blocks. An example is shown in Fig.2 where a canal bridge is collapsing and the blocks have completely separated. A bridge constructed with modern stronger and more adhesive mortar would have collapsed by a different mechanism. This bridge was under test and reached a maximum load of 211 tonne positioned across a line at ⅓-span, Page (1987). The weak mortar had enabled the structure to cope with temperature effects over the years without any serious loss in load carrying capacity.

Fig.2. Collapse of masonry arch bridge.

## 2 Iron bridges 1750 - 1890

Iron was introduced into bridge construction in the latter half of the eighteenth century, Ironbridge (the oldest survivor) being constructed across the Severn at Coalbrookdale in 1777. The early structures were a development of the arch form with no provision for articulation. The ends of the ribs were fitted into grooved sockets in the abutment bearing castings and were considered to be structurally fixed. Temperature effects became apparent at an early stage and during construction of Southwark bridge in 1818 it was found that the average effect of expansion caused by summer increases in temperature was a rise in height of the crown of 38mm. It seems that iron arch bridges were able to deflect without provision for expansion and without causing undue distress to the abutments in the short term if not the long term.

The introduction of iron enabled designers to turn to suspended spans of an altogether more flexible nature. Notable examples that are still in service include Union bridge across the Tweed (1820) and the Menai bridge (1826). These early examples were troubled by wind induced vibration and there were a number of bridges that were damaged or destroyed by wind. Designers were therefore forced to focus their attention on improving the dynamic stability of suspension systems and temperature effects were of secondary importance if they were considered at all. James Dredge of Bath devised a patented system of cable configuration which evidently worked and several of his bridges are still standing, for example Victoria bridge across the Avon at Bath (1836). There was no provision for expansion in Victoria bridge, the main beams being very lightweight and fixed into the masonry piers with a mortared joint. When inspected recently it was evident that relative movements occur and the joints have converted from a fixed to a sliding action.

Robert Stephenson's Britannia bridge, constructed in 1846, as well as being one of the most researched designs, was an early example of the use of roller bearings. The box girder is fixed at the central Britannia tower but free to slide on cast iron roller bearings through the other towers, Drysdale Dempsey (1864). Each of the half lengths is 461m. Using a coefficient of expansion measured by Professor Daniell in 1831, it was calculated that the required temperature movement was 152mm. An expansion of 82mm was measured for a temperature increase of 14°C. During construction it was recorded that differential temperature effects caused the box to bend 57mm laterally and 63mm vertically.

An awareness of temperature effects was expressed in relation to the first Tay bridge (1878). The Board of Trade Inspector, Major General Charles Scrope Hutchinson, recorded in his acceptance report that expansion effects in hot weather should be reduced as much as possible by painting the girders white. He also wrote that he would visit the bridge again to observe the effects of a high wind when a train of carriages ran across. In the event the Tay bridge was destroyed by a gale before he returned.

During the railway era some 25000 bridges were built between 1830 and 1860 and the state of bridge engineering was developed very considerably. The requirement to avoid steep inclines forced designers to build bridges and viaducts of lengths and heights that were greater than anything since the Roman period. Due to the dynamic effects of steam engines, suspension bridges of the time were too lively and designers turned to other structural forms, such as through trusses, when long spans were required. A case in point is the Hungerford bridge, across the Thames in London, which replaced Brunel's suspension bridge for pedestrians. It was reconstructed as a 6-span structure incorporating the original towers and having a continuous truss. The designer apparently failed to make due allowance for the tensile stresses developed by negative bending over the piers and the top flanges cracked; it is not known when the cracking occurred but it did not appear to cause any problems because the bridge performed satisfactorily until increased loading led to the need for extensive strengthening in 1979.

Fig.3. Fractured iron work over pier of Hungerford Bridge.

## 3 Concrete and steel 1890 - 1950

Concrete was introduced towards the end of the nineteenth century, for example, Glenfinnan Viaduct which was built in 1897. The structure was composed of twenty one spans of 15m at heights up to 27m. It was considered that there was a possibility of differential settlement; this was catered for by having 12mm steel plate sliding joints at the crown of each arch. There was no provision for longitudinal expansion.

The first bridge at Angel Road on the A406, built in 1908, was composed of a reinforced concrete deck on continuous beams. It had 17 equal spans and a total length of 223m. The structure evidently performed satisfactorily until replaced in 1960.

Wadell (1916), advised against the unnecessary use of expansion joints because they could weaken the structure, cause 'rough spots' on the running surface, lower the structural rigidity and be more expensive. Arches and suspension bridges did not require provision for expansion as they could deflect. However joints would always be necessary when girdere were fixed to masonry. For concrete structures he advised that joints should be provided every 15m to avoid cracking.

In Arrol's handbook of steel bridges and structures, Hunter (1920), information about temperature effects on steel structures was provided. Temperature data collected over the previous 20 years were quoted and measurements of bridge expansions provided. In a section giving recommendations for design of steel railway bridges, it was recommended that provision be made for expansion and contraction at the rate of 25mm per 30m. In relation to the girders it was stated that suitable heavy plates should be fixed at one end and expansion bearings provided at the other. It was also noted that there should be restraint of transverse motion and uplift.

Bressey (1933), the then Chief Engineer of the Roads Department of the Ministry of Transport, writing on bridges in general, dealt with effects of differential settlement and mentioned that cast iron arches could have three hinges to cope with moderate settlement and cited the Sunderland Wearmouth bridge which was above old mine workings. Steel bridges of the time usually had hinges at the springing points but these were designed to be for ease of construction. Reinforced concrete bridges were mostly without hinges but temporary ones were occasionally used during construction. Expansive effects due to temperatures were not mentioned,

Bressey noted that if multi-spans were made continuous over the piers, bending moments would be reduced at the centre of the spans and less material would be needed. The same argument would apply to single spans fixed to the abutments, the longest span portal at that time being Wisbech bridge which was 28m.

Chettoe and Adams (1938) writing about design of concrete bridges, noted that the provision for expansion and contraction of reinforced concrete superstructures is frequently ignored, sometimes with unhappy results. The problems usually occurred due to the abutments and wing walls being pushed backwards and forwards. Although admitting inadequate knowledge about the extent of the movements, and the length of span when expansion becomes important, it was recommended that expansion joints are not generally needed for spans of up to 9m. For reinforced concrete a temperature range of 33°C was suggested as being reasonable and took into account the fact that the full range of air temperatures would not be experienced by the superstructure due to the low thermal conductivity of the concrete. This would give a movement of 11mm per 30m length. Contemporary practice was to fit joints at the ends of bridges but for bridges longer than 42m joints were sometimes positioned over the piers.

Chettoe and Adams (1938) outlined the types of expansion joint and bearing then being used: sliding plates, simple cast iron rockers, heavy cast iron or cast steel rockers, rollers, reinforced concrete rockers.

## 4 Recent developments 1950 - present

The use of continuous superstructures recommended by Bressey was not generally adopted possibly because of factors such as the uncertainity of differential settlement and the greater economy of using standard beams with simply supported spans. After the halt in new construction during the 1940's the motorway era starting in the late 1950's contained examples of long viaducts having simply supported spans, for example the Midland Links. There were also some notable continuous structures, for example Westway and Hammersmith.

In the 1970's more comprehensive measurements of bridge temperatures were made for different types of bridges by the Transport and Road Research Laboratory and it became evident that ranges of temperature are much greater than hitherto considered. This produced higher calculated values of expansion and further discouraged design of long continuous spans.

With the development of higher strength concrete and mortars there has been a loss in the ductility which enabled earlier structures to cope with thermal effects. This is well illustrated by brick and stone structures which until comparatively recent times did not require expansion joints. In contrast design of brickwork now requires 20mm joints at 10m intervals containing a filler capable of compressing to 10mm. This procedure is even adopted for non structural brickwork such as garden walls. The efficacy of waterproofing the joints is dependent on the filler adhering to the brickwork; in practice this is a common source of failure and there are examples of splitting at the interface after very short times.

In highway bridges failures of expansion joints became evident at an early stage but for some time the resulting leakage was considered to be only a nuisance permitting unsightly stains to develop. In the time when gritting was used to combat the effects of ice and snow there were no significant problems but this changed after the introduction of deicing salt in the mid-1950's. The first signs of corrosion problems were in decks, particularly in the United States, when it was found that chlorides from deicing salt were corroding the steel reinforcement. A few years later it became evident that leakage through expansion joints was, depending on the design of the bridge in question, permitting salt water to flow over the substructure causing corrosion to the reinforcing steel in crossheads and piers. Less commonly there was corrosion of anchorages and tendons of post tensioned beams. In a survey of the performance of concrete in 200 of the Department of Transport's bridges, Wallbank (1989) reported that three quarters of all joints were leaking.

The problems caused by leakage has focused attention on the need to develop more durable designs of expansion joints, particularly in relation to their ability to resist the action of fatigue and remain watertight. In this context several investigations are now being conducted to determine fatigue performances and improve designs.

A very different approach is to design continuous bridges to avoid the need for expansion joints over the piers. There are two categories of continuity; continuous superstructures having expansion joints at their ends and integral bridges having superstructures that are either fixed to their abutments as in portal frames or free to slide so that the end of the deck is effectively absorbed into the pavement. Modern jointless bridges, as opposed to the earlier types built in ignorance, have been constructed in the United States in the last 20 years, particularly in Tennessee. To date the performance of these bridges has been said to be very good and there has been no serious cracking or degradation reported. It is notable that in this time there would almost certainly have been failures of expansion joints. As with the earlier masonry viaducts, integral bridges appear to defy the laws of physics. There are several possible mechanisms that may operate to permit integral bridges to work:

- Some of the movement may be transferred to a rocking action on the abutments.
- The full temperature range is usually experienced over a relatively long period of time and this would enable the structure to creep.
- The structure may be able to cope with the more rapid ranges of temperature by distortion.

In the long term jointless bridges constructed with modern less forgiving materials are liable to exhibit distress through fatigue cracking of the concrete but this is likely to be in times comparable with the design life. This is exemplified by recent works to refurbish a 60 year old 4-span bridge having a total length of 65m. The masonry piers and abutments were from an earlier bridge constructed in 1875. The deck was of continuous reinforced beam-and-slab construction and no provision had been made for expansion. Over the years the masonry abutments became cracked and deteriorated to the extent that extensive repairs were required. Nevertheless, it had performed satisfactorily without the problems posed by expansion joints.

## 5 Concluding remarks

In the early types of structure it was possible to ignore effects of expansion because construction materials were more ductile and forgiving, Moreover, components were smaller so that relative movements and multi-hinges could form enabling the structure to deform. Modern high strength materials enable longer spans and lighter weights to be achieved but structures are less able to deform.

It was proved virtually impossible to devise expansion joints that remain serviceable and water tight for a reasonable period of time. In consequence the maintenance costs of modern bridges are increased due to the cost of repair works and consequential delays to traffic.

Experience of reinforced concrete bridges constructed in the early years of this century has shown that jointless construction can last for 60 years or more. In cases where damage occurred it appears to have been due mainly to fatigue processes. Such damage may have been less expensive to repair than articulated bridges having failed expansion joints and consequential corrosion.

Experience in the United States, over the last 20 years indicates that it is possible to operate long span concrete bridges having no expansion joints, without apparent distress.

It appears that current allowances for temperature effects may be excessive and there is scope to reconsider methods of designing to cope with temperature effects.

## 6 References

Bressey, C.H. (1933) The Development of Road Bridges in Great Britain. British Bridges, pp vii-xxiv, Public Works, Roads and Transport Congress.
Chettoe, C.S. and Adam, H.C. (1938) Concrete Bridge Design.
Drysdale Dempsey, G. Tubular Bridges, (1864). Virtue Brothers & Co.
Hunter A. (1920) Bridge and Structural Engineering Handbook, Sir William Arrol, Spon.
Page, J. (1987) Load tests to collapse on two arch bridges at Preston, Shropshire and Prestwood, Staffordshire. Department of Transport TRRL Report, RR110, TRRL Crowthorne.
Shirley Smith, H. (1954) The Worlds Great Bridges. Phoenix.
Waddell. (1916) Bridge Engineering. Chapman and Hall.
Wallbank, E.J. (1989) The Performance of Concrete Bridges, HMSO.

# 6 ADVANTAGES OF BRIDGE DECK CONTINUITY

B.P. PRITCHARD
Consultant to W.S. Atkins Consultants Ltd and Colebrand Ltd, Haywards Heath, UK

**Abstract**
Multispan bridge decks which are designed to provide continuity across the support piers offer a number of advantages, both in the long-term performance of the structure, but also in simplification of the several aspects of construction, and reduced overall construction costs. This paper outlines these advantages and gives examples of comparative simply supported and continuous designs.
Keywords: Deck continuity; reduction of joints; reduction in depth of flyovers and underpasses; reduction in length and height of flyover and approach ramps; reduction of bearings; construction depth; pier and foundation savings; security of extra redundancy.

This paper has been prepared from Chapter 2 of the author's book 'Bridge Design for Economy & Durability', published in 1992. Thanks are due to the publishers, Thomas Telford Publications, for permission to reproduce it. The book is available from the Thomas Telford Bookshop, ICE, London or from Thomas Telford Services Ltd, Thomas Telford House, 1 Heron Quay, London E14 4JD.

## 1 Deck continuity

Early multi-span beam and slab or slab bridges were generally designed and built as a sequence of simply supported spans. In 1930 the American, Hardy Cross, introduced his famous moment distribution method for the quick and easy analysis of continuous frames and beams[1]. This method was immediately taken on by bridge designers and a gradual build-up in the use of deck continuity commenced.

The 1950s and 1960s saw a further great improvement in continuous-deck analysis with the introduction of computers, together with grillage and finite element programs. The use of continuity spread, particularly in winter-affected countries where the maintenance problems of deck joints and de-icing salt have dramatically multiplied. In some countries national bridge codes recommend or even dictate multi-span deck continuity unless ground settlement problems require the greater articulation provided by a series of simply supported spans.

Currently, the UK Department of Transport is considering the issue of an Advice Note encouraging bridge designers to use deck continuity, principally to

*Continuous and Integral Bridges*. Edited by B.P. Pritchard. Published in 1994
by Taylor & Francis, 2 Park Square, Milton Park, Abingdon, Oxon, OX14 4RN. ISBN: 0 419 19030 9.

Fig. 1 A5 flyover, Staples Corner Interchange, London. The joint-free deck acts as an umbrella over the pier.

minimise the use of trouble-prone deck joints. The note will also recommend 'integral bridges', in which the continuity of bridge decks free of intermediate joints is extended to the elimination of the end expansion joints by building the deck into the abutments.[2]

The elimination of deck joints is a good and sufficient reason for adopting deck continuity, with all the implied savings in maintenance costs. However, it should be remembered that deck continuity can also offer considerable economies in first-time construction costs. It is this aspect which is emphasised in sections 2 and 3.

Continuity can be adopted for most multi-span bridges. Only in very short span structures and in mining areas will differential pier and abutment settlements become significant enough to dictate a change to simply supported spans.

Differential settlements in highway bridges rarely exceed 10–20 mm and in the more critical shorter spans there is a tendency to use reinforced concrete decks anyway. Because of the lower concrete elasticity modulus and inherent cracking, reinforced concrete is less stiff than prestressed concrete and accommodates differential settlement with less attracted moment.

In certain cases single-span or three-span decks subject to significant settlements can be given some of the benefits of continuity by counterweighting[3].

## 2 Advantages

Figures 1 & 2 illustrate the main advantages to be gained by using multi-span deck continuity: Fig. 2(a) shows a typical simply supported multi-span deck, Fig. 2(b) a corresponding continuous deck.

The first advantage is the elimination of deck and parapet joints at the piers. This represents savings in both construction and maintenance costs - maintenance which includes that of the underlying piers. Elimination of joints also gives the vehicles crossing the bridge a better ride.

The second advantage is the saving in deck depth, which not only can reduce the construction cost of both the deck and the supporting substructure but also can reduce the length of the deck or the associated approach ramps or earthworks. The latter saving is probably the most significant construction-cost economy and is dealt with separately in section 3 below.

The third advantage lies in the halving of the number of bearings required at each pier. Although the bearings will be of higher capacity, the lesser number should result in cost savings.

The fourth advantage results from the replacement of the two rows of bearings by the single row of central bearings. The piers or crossheads can be

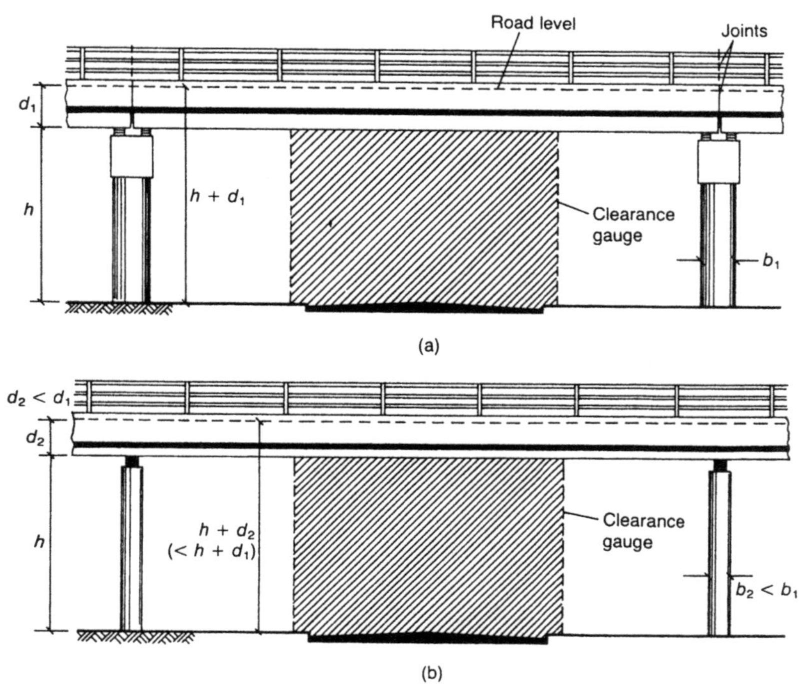


Fig. 2 Advantages of continuity: (a) simply supported deck; (b) continuous deck.

reduced in thickness not only because the single row of bearings takes up less room at the top, but because the deck dead load has reduced and the live load moments applied by off-centre pairs of bearings are removed. Significant cost savings in pier foundations also result.

A fifth advantage, possible only with deck continuity, is that the 'sucker deck' principle can be applied to advantage[3].

A sixth advantage, again possible only with deck continuity, is that use can be made of the transverse strength of the deck over piers (equal to the longitudinal strength for slab decks) to provide inboard piers and single bearing support. Considerable benefits can result from simplification of skew crossings and land savings by passing interchange slip roads under the main flyover[4].

A seventh advantage is that continuity means extra deck redundancy, which generally guards against sudden collapse in overload situations or in instances of sabotage (Fig. 3).

Fig. 3 Bomb damage to A5 flyover, Staples Corner Interchange.

## 3 Savings in deck construction depth

The optimum construction depth/span ratios vary with span, live loading, materials and methods of construction. Average ratios for the 10–60 m span range are about 1/24 for continuous spans and 1/18 for simply supported spans. Thus, for 35–40 m spans about a half metre can be saved in deck construction depth by adopting continuity. In itself, the reduction in concrete or steel would appear to lead to a reduction in construction cost. This is not always the case, as continuity can introduce extra complications to the sequence and standardisation of construction, usually more perceived than real if design and construction are adequately planned to suit the continuity.

Whether the continuous deck is cheaper to build or not, the associated saving in construction depth can certainly reduce construction costs of other elements of the bridge or viaduct. The substructure, of course, can benefit by reductions of up to 10–15% in deck weight, significant with the longer spans and where foundations are poor. However, it is the possible reductions in related earthworks, approach ramps or bridge lengths which are most cost-effective. Typical examples are demonstrated in Figs 4–6.

Figure 4 shows a typical urban flyover carrying a highway over a ground-level roundabout. The approach ramps rise at a 4% grade and the 180 m long viaduct has a compliant vertical curve providing the required headroom over the underlying roundabout carriageways. In Fig. 4(a) there are five equal 36 m simply supported spans with a deck construction depth of 2 m. The resulting maximum ramp height is 7 m, giving an overall length of 175 m for each ramp.

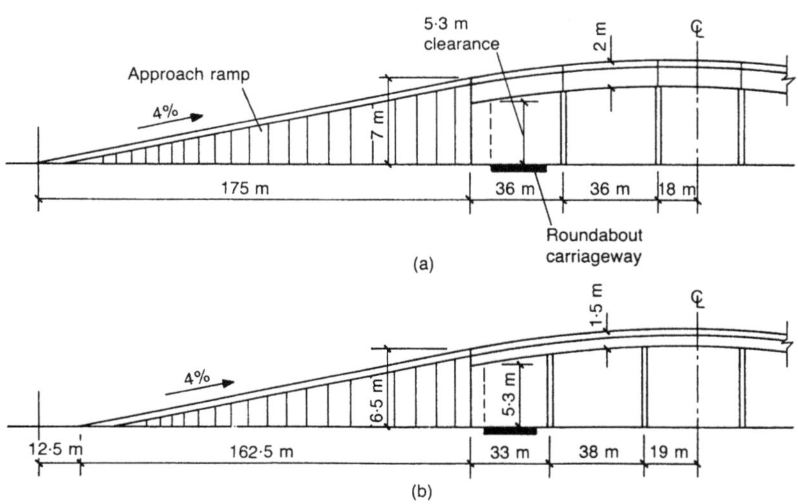


Fig. 4 Approach ramp savings: (a) simply supported spans; (b) continuous spans

Fig. 5 M1 flyover, Hendon Fiveways Interchange, London.

The proposal in Fig. 4(b) uses span continuity and slightly modifies the span layouts to give the more balanced arrangement of shorter end spans. This results in three inner spans of 38 m and two end spans of 33 m. The associated reduction in deck depth of 0.5 m means that the maximum ramp height drops to 6.5 m and the ramp length reduces to 162.5 m.

The total reduction in the approach ramps due to the introduction of viaduct continuity is therefore measured by a 7% reduction in length and height. This results in a reduction of nearly 14% in ramp wall area, which also gives a rough approximation of the cost saving in ramp construction at each end. Fig. 5 shows an early flyover over the M1 which benefited with significant ramp savings generated by the depth reduction in the continuous decks.

Figure 6 shows another flyover with a similar vertical profile, this time crossing a centrally located urban carriageway flanked by pedestrian areas. In this case, the designer again chooses to use approach ramps rather than embankments to reduce landtake. His calculations show that the ramps can rise to a height of 8 m using spread footings for the ramp walls, while remaining cheaper per square metre than suspended deck. Anything higher would require expensive piling, exceeding the cost of suspended deck. The maximum length of each of the 4% ramps is therefore set at 200 m.

A 36 m span is required across the central carriageway and Fig. 6(a) shows a sequence of three equal 36 m simply supported spans, the outer spans crossing

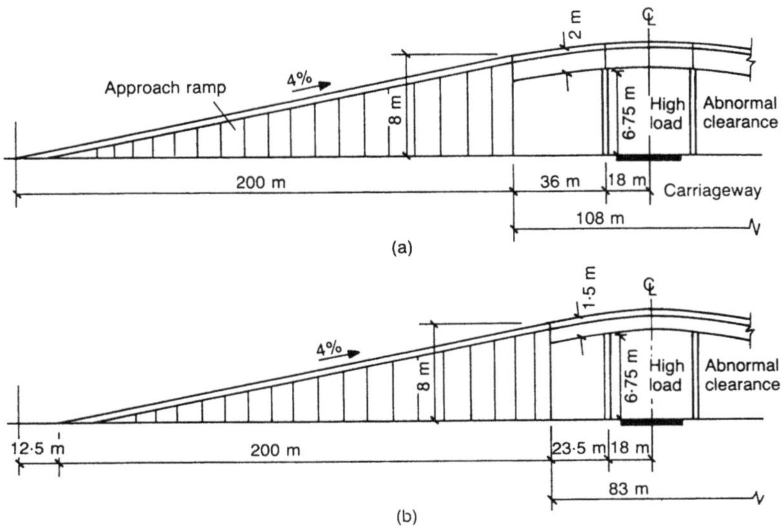


Fig. 6 Savings in flyover length: (a) simply supported spans; (b) continuous spans.

the paved pedestrian areas. The 2 m deck construction depth results in the maximum allowable ramp height of 8 m and length of 200 m.

Fig. 6(b) shows the effect of providing span continuity and the associated 0.5 m saving in deck construction depth. The 200 m long, 8 m high ramps are retained, but each can now be relocated 12.5 m nearer the central carriageway, reducing the required viaduct length from 108 m to 83 m and the end spans from 36 m to 23.5 m. Thus, with a small loss of pedestrian area and no change in the approach ramps, the flyover viaduct length is reduced by some 24%, with similar percentage savings in viaduct cost and, possibly, landtake. For a dual three-lane motorway flyover, current deck costs would indicate a saving of some £400,000–£500,000.

The final example, shown in Fig. 7, is a dual three-lane motorway locally depressed to pass under twin 36 m span simply supported bridges carrying a 160 m diameter ground level roundabout. In this case the vertical profile of the motorway consists of an 80 m long crest curve followed by a 4% downgrade section 150 m long, then a sag curve 160 m long followed by 150 m of 4% upgrade and a final 80 m long crest curve. The bridges are located just above the ends of the sag curve, which are 7.3 m below ground level. This depth provides the required 5.3 m clearance under the 2 m deck construction depth.

The depressed section of motorway is 620 m long and the excavation required with 1 : 2 cutting slopes will be around 150,000 m$^3$. Savings can be made by adding continuity to the roundabout bridge decks. The 0.5 m deck construction depth saved allows the depressed motorway profile to be raised by the same amount, shown as a broken line on the figure. This saves approximately 15,000 m$^3$,

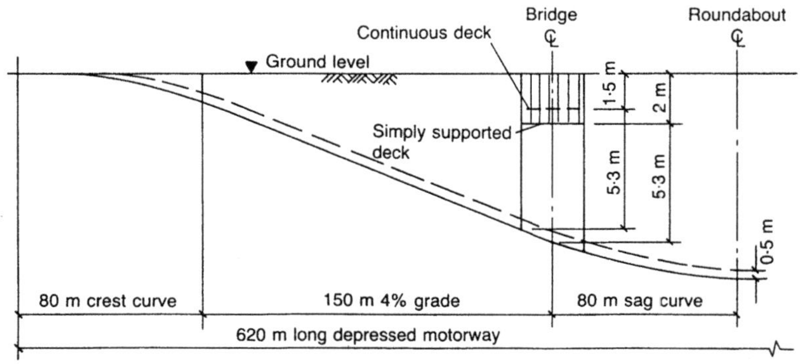


Fig. 7 Excavation savings, depressed motorway.

or 10%, of the excavation. There are also small savings in land acquisition and, indeed, in the overall length of the roundabout overbridges.

Depending on the excavated material and its ease of disposal, the excavation cost savings which result from changing the roundabout bridge deck from simply supported to continuous can readily exceed £100,000. In rock, of course, the saving could be many times higher. Fig. 8 shows such a depressed interchange, in this case with an additional top-layer flyover.

Fig. 8 Depressed M25 Interchange.

## 1 References

1 Cross, H. Analysis of continuous frames by distributing fixed-end moments. *Proc. Am. Soc. Civ. Engrs, 1930.*

2 Burke, M. The integrated construction and conversion of single and multispan bridges. *Bridge Management*, Proceedings of International Symposium, Surrey University, 1990. Elsevier.

3 Pritchard, B. P., Haunched bridges and the Sucker Deck Principle, *Proceedings of the Institute of Civil Engineers, Structures and Buildings*, February 1994.

4 Pritchard, B. P. *Bridge design for economy and durability*. Thomas Telford, London, 1991.

# 7 CONTINUITY IN DECKS WITH PRECAST BEAMS – PRACTICAL ISSUES

H.P.J. TAYLOR
Costain Building Products Ltd, Stamford, UK

**Abstract**
This paper considers the design issues involved in making fully continuous decks from simply supported bridge beam arrangements and the practical considerations that they impose.
**Keywords** Continuity, design, limit states construction joints.

## 1 Introduction

Factory made pretensioned prestressed concrete bridge beams have been one of the success stories in bridges over the last 40 years. Particularly in the UK and the USA many thousand bridge decks using these beams, as infil slab decks, beam and slab decks and voided decks, have been constructed with remarkably few performance difficulties with respect to the beams and decks. The main area of lack of good durability has been in the weakness of the various forms of joints in what are usually simply supported decks. The obvious answer to this is continuity of some form.

Continuity in bridge decks with pretensioned precast beams is not new, a significant amount of research and design development data exists in the literature, dating back to the early 1960's.

## 2 Practical Issues

From the point of view of ease of constructuction and economy, the most suitable method of constructing a continuous joint is to maximise the precast element and minimise the site work and the number of bearings, Fig.1(a).

The narrow insitu connection over a single line of bearings produces the following advantages.

- Bearing numbers are minimised, bearings are expensive
- Beams and insitu shutter can all be supported from the main pile cap
- Insitu concrete is minimised
- Shuttering is minimised
- The method is appropriate for wall or column support

*Continuous and Integral Bridges*. Edited by B.P. Pritchard. Published in 1994 by Taylor & Francis, 2 Park Square, Milton Park, Abingdon, Oxon, OX14 4RN. ISBN: 0 419 19030 9.

This form of construction requires that two features are present. The spans should be such that the precast beams are within the transportable range and the design should be such that there is not too much congestion of reinforcement in the insitu joint. The use of two lines of bearings seems to be unnecessary Fig.1(b)

In this case, movement of the deck as the loaded length changes from one span to another may require that each bearing be able to carry more than half the load for a single line, some designers make each line capable of carrying the full load with a consequential increase of cost.

Wider joints may be used and may be appropriate with skew decks or with a bridge curving in plan. In this case it is important to consider the relative stiffness of the support of the beams and the insitu concrete shutter. Settlement during the casting operation could result in plastic cracks in the insitu concrete and in extreme cases, cracks at the top of the precast beams. Shutter support in the form shown in Fig.2(b) should be avoided, the ideal method is to use the form shown in Fig.2(a).

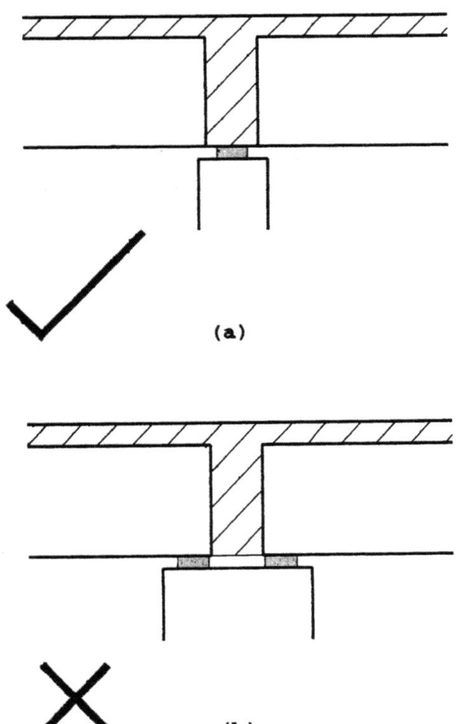

(a)

(b)

**Fig. 1. Minimizing the number of bearings**

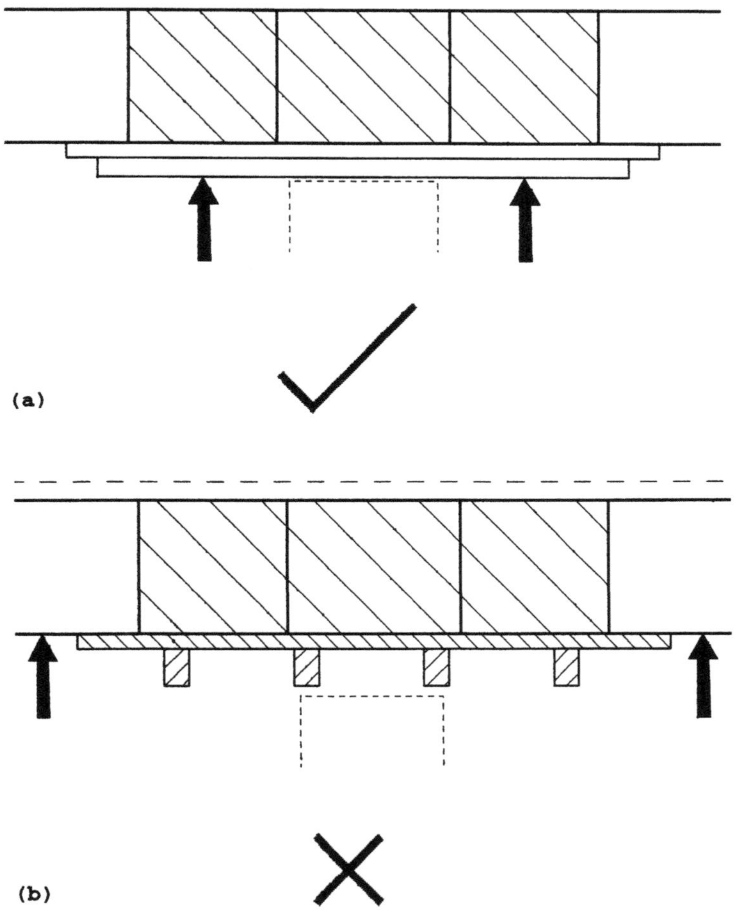

(a)

(b)

**Fig. 2. Shutter support for wider beams.**

### 3 Design Considerations

The integral joint over a support consists of precast prestressed beams with a reinforced concrete slab and infil diaphragm. This creates a hybrid section which may present some confusion in analysis.

**SLS Hogging Bending**

At the serviceability limit state (SLS) the section may be subjected to hogging moments Fig.3. In this case, in BS5400 the top slab is considered to be reinforced, with a crack width criterion and the prestressed beam prestressed with a no tension criterion. The designer has to decide whether cracks in the slab may be

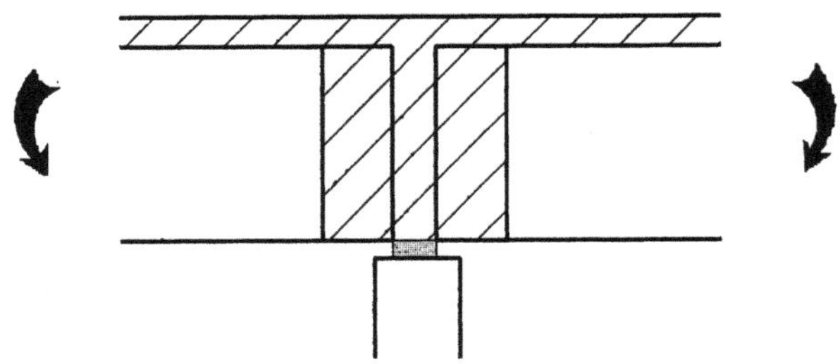

**Fig. 3. Hogging at SLS and ULS.**

permitted to progress down into the top of the prestressed beam at the SLS. Such cracking cannot be important from the durability point of view as it would not pass through any of the prestressed strands. Research on composite construction has shown that at a joint between precast and insitu concrete the tensile cracking strain and cracking behavior of the weaker section is controlled by the stronger. Design studies and a review of the literature on composite construction are required in this area in order that positive advice may be given to designers, within the requirements of BS5400.

**ULS Hogging Bending**

A further consideration of the same section is hogging bending at the ULS. BS5400 contains methods for the calculation of the ultimate strength of sections which can take account of stressed and unstressed reinforcement in both the tensile and compressive zone. It is possible to modify the concrete stress block to take account of the pre compression on the bottom flange of a prestressed beam but designers are unfamiliar with the concept. The strain compatibility method, illustrated in Fig.4 is well known and the modifications shown in the figure to modify the stress block should be developed and compared with the literature. The method, when used to calculate the strength of the section tested by Regan,[1] shows good fit.

**Parasitic Effects**

The prestressed beams will continue to deform due to long term creep and differential shrinkage after the deck is made continuous. This effect will produce parasitic moments, as shown in Fig.5 and in the set of bending moment diagrams in Fig.6. This figure gives a summary of the moments produced in an actual case of a three span bridge and shows that the effect is significant and can go as far as producing net sagging moments at the support in the unloaded case. This effect has also been studied in the USA in the early 1960's[2]. The reinforcement details for this can be difficult to contain within a narrow joint.

An alternative to this is not to attempt to reinforce but to allow the joint to crack in this situation, relieving the parasitic moment but producing a crack extending from the support, upwards towards the top of the bridge[3]. Such a crack does not expose critical elements

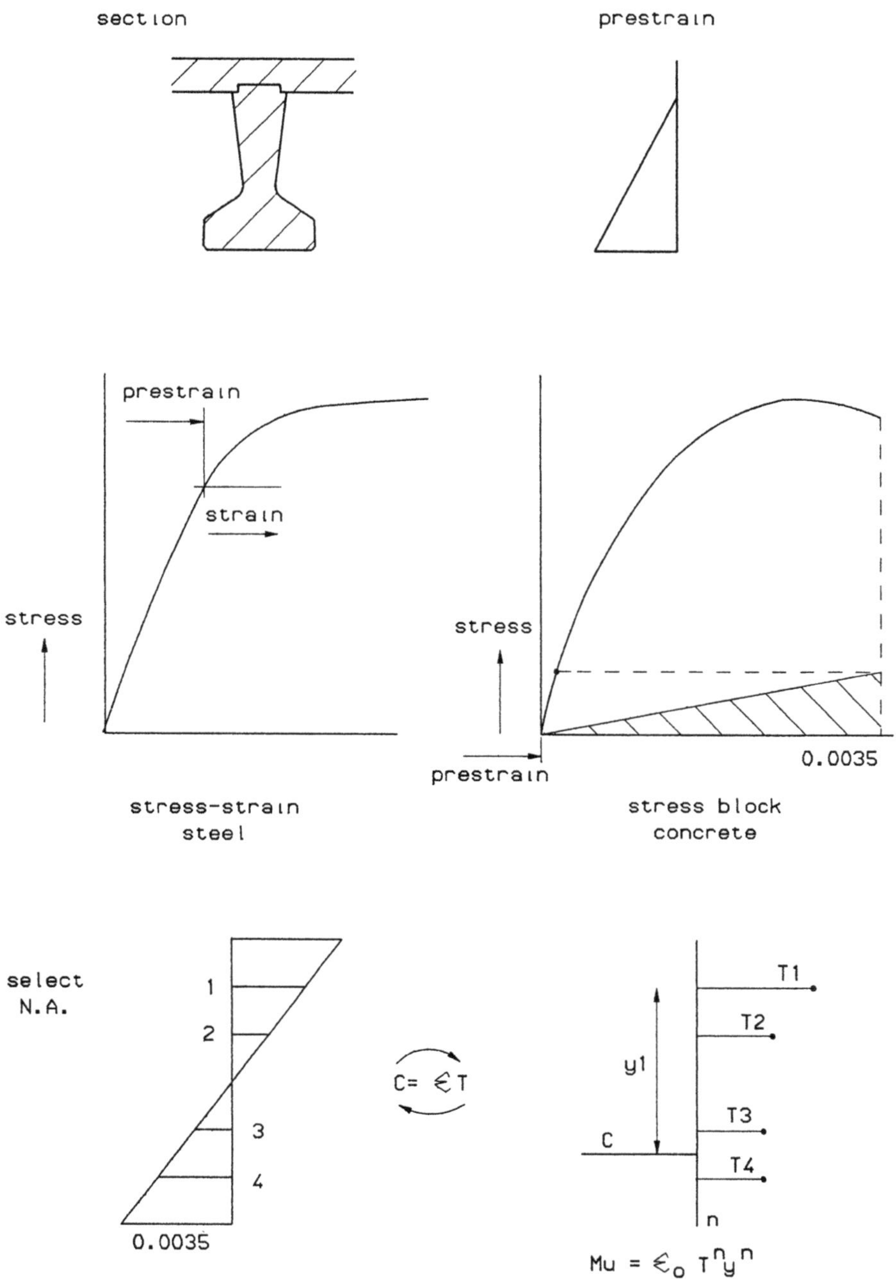


**Fig. 4. Strain compatibility method accounting for prestrained concrete and steel.**

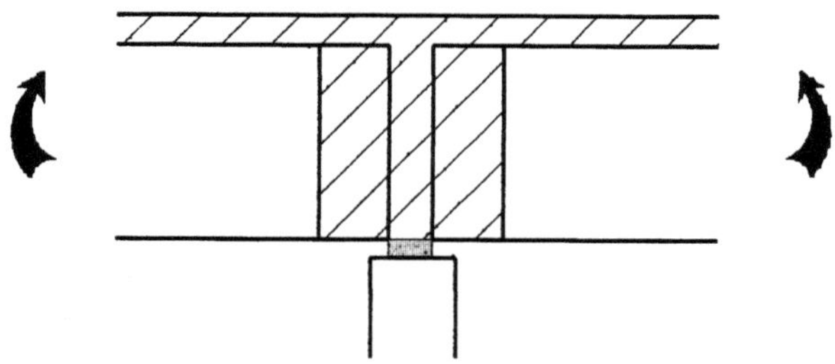

**Fig. 5. Possible sagging at SLS.**

of the joint to ingress of water and salt from above and there is a large body of evidence to show that it does not produce a durability problems.

This subject should be reviewed and researched to produce sound advice within the permitted extent of BS5400 and should be developed to a simple design process that can easily be fitted in to current computer systems for grillage and finite element analysis.

## References

**Regan, P.E.** Behaviour of Precast Prestressed Y-beams in Shear, Torsion and Negative Bending Structural Engineer Vol 68 No 23 December 1990

**Karr, P.H., Kriz, L.B.** and **Hognestad, E.** Precast Prestressed Concrete Bridges Pilot Tests on Continuous Girders Development Bulletin D34 PCA Illiniois May 1960

**Hambly, E.** and **Nicholson, B.** Precast Integral Bridges Prestressed Concrete Association Leicester 1990

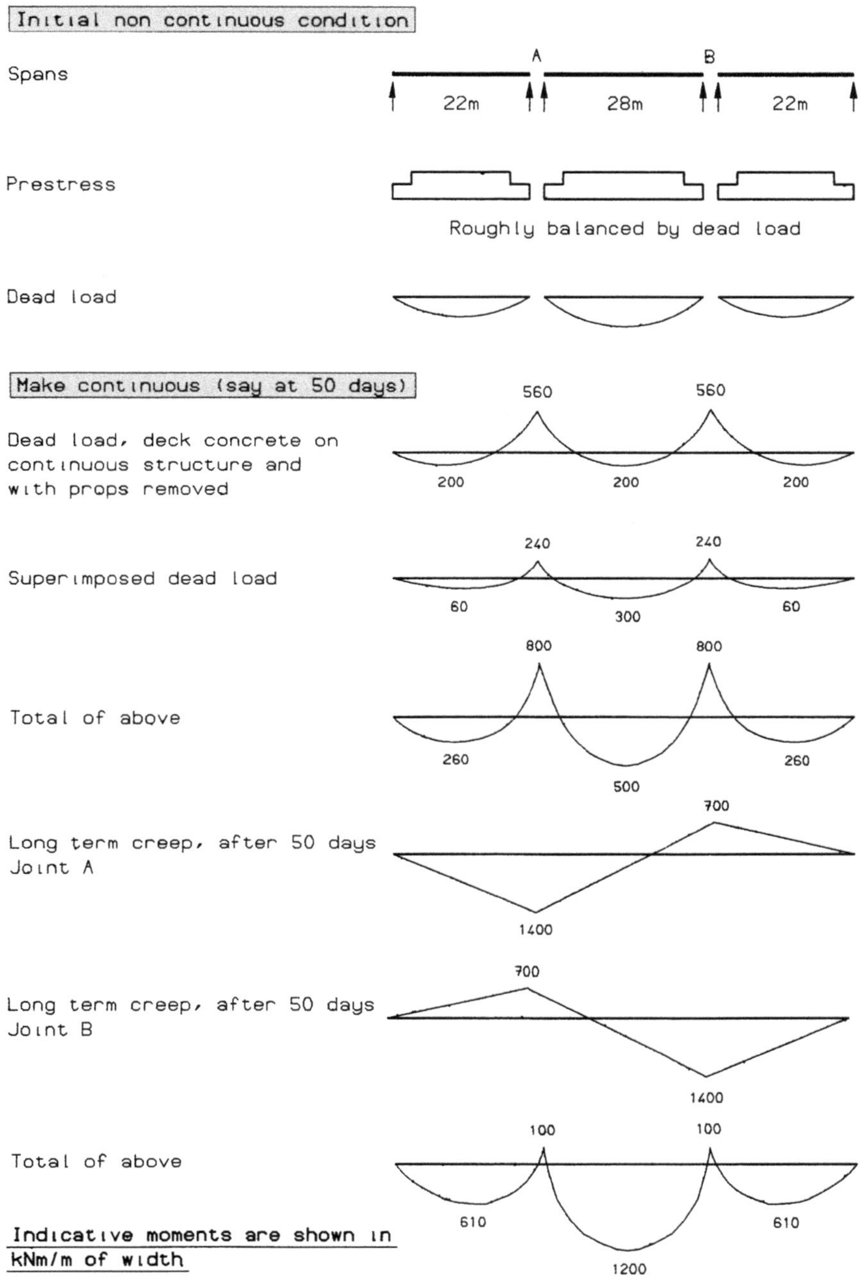


**Fig. 6. Stages in design of a typical precast, prestressed beam bridge, made continuous.**

# REPORT ON SESSION TWO

S. SHANMUGAM

Dr Tilly presented an interesting account of the history of bridge continuity and showed several examples of early bridges dating from 1600 BC built in stone, brick masonry, iron, concrete and steel to the more modern designs post 1950. He went on to explain the possible mechanisms outlined in his paper together with the reasons why such early bridges were able to cope with the temperature effects. He considered that there was a need for a review of the current allowances being made for temperature effects in the light of satisfactory performance of so many jointless bridges in recent years.

In introducing his paper, Mr Pritchard showed several stark examples of bad details such as leaking joints, corroded bearings, and deterioration of pier cross heads, arising from the lack of bridge continuity. He highlighted many inherent advantages of multi-span deck continuity by showing actual continuous span layouts which had resulted in large savings in both construction and maintenance costs. He pointed out that the extra deck redundancy in continuous structures would also guard against sudden collapse in the event of sabotage, citing an example where bomb blast had lifted the deck of a flyover by 200 mm and blown holes in the box flanges.

Dr Taylor explained the issues involved in making decks fully continuous from simply supported bridge beam arrangements. He described research programmes carried out notably by Karr, Kriz and Hognestad on pilot tests on continuous girders, and by Regan on the behaviour of precast prestressed Y beams in shear, torsion and negative bending.

Dr Taylor illustrated several examples of diaphragm layout with the most beneficial ways of supporting the precast beams and the in situ diaphragm shutter; he explained how the strength of the reinforced support section in negative bending at Ultimate Limit State, sagging moment at the support under long term creep of the beams, and hogging moment over the support at Serviceability Limit State could be adequately dealt with in practice.

# PART THREE

# CONTINUOUS BRIDGES

# 8 DESIGN CONCEPT OF THE TRANS-TOKYO BAY BRIDGE

Y. SHIOI
Design and Engineering Department, Trans-Tokyo Bay Highway Corporation, Japan
S. NAKAMURA
Nippon Steel Corporation, Kobe, Japan
(formerly Trans-Tokyo Bay Highway Corporation)

Abstract
This paper intruduces the design concept of the Trans–Tokyo Bay Bridge, a group of continuous girder bridges. Structural forms of the super–structures, sub–structures, bearings and expansion joints are also presented.
Keywords: Continous girder bridges, Rubber bearings

## 1. Outline of the Trans–Tokyo Bay Bridge

The Trans–Tokyo Bay Highway (TTBH) is a 15.1km long toll highway connecting Kawasaki and Kisarazu which are situated at both sides of the Tokyo Bay. The highway consists of two 10km long tunnels under the Kawasaki waters where marine traffic is heavy, and a 4.5km long bridge over the Kisarazu waters where marine traffic is relatively sparse, as shown in fig.1. Two man–made islands are also situated at the middle of the tunnel and at the transition point of the tunnel and the bridge.

It should be noted that, although most of the public highways in Japan have been constructed by public corporations, a semi–public company was founded for construction of the TTBH in order to utilize financial and human resources of both public and private bodies.

The highway carries dual two lane carriageway in the first stage, and other two lanes are planned to be extended in the future depending on the future traffic flow. Traffic volume is estimated approximately 33,000 veh/day which is expected to reach 64,000 veh/day in 20 years after opening. Design vehicle speed is 80km/h.

The bridge is 4.4km long with 42 piers. The navigation area exists at water depth about 20m, where the span is expanded up to 240m. The layout of the bridge is shown in fig.2 and fig.1, in which the soil layer is also illustrated. Alluvial sand and clay layers (A) exist 5m to 20m deep under the seabed. In some alluvial layers sands are soft with N–value of the standard penetration tests under 10 and the grains are uniform and medium size, which might create a liquefaction problem. Below the alluvial soils follow the diluvial soil layers with different ages, D1 to D4, where sands and clays exist alternately. D5 layer observed overall the

*Continuous and Integral Bridges*. Edited by B.P. Pritchard. Published in 1994
by Taylor & Francis, 2 Park Square, Milton Park, Abingdon, Oxon, OX14 4RN. ISBN: 0 419 19030 9.

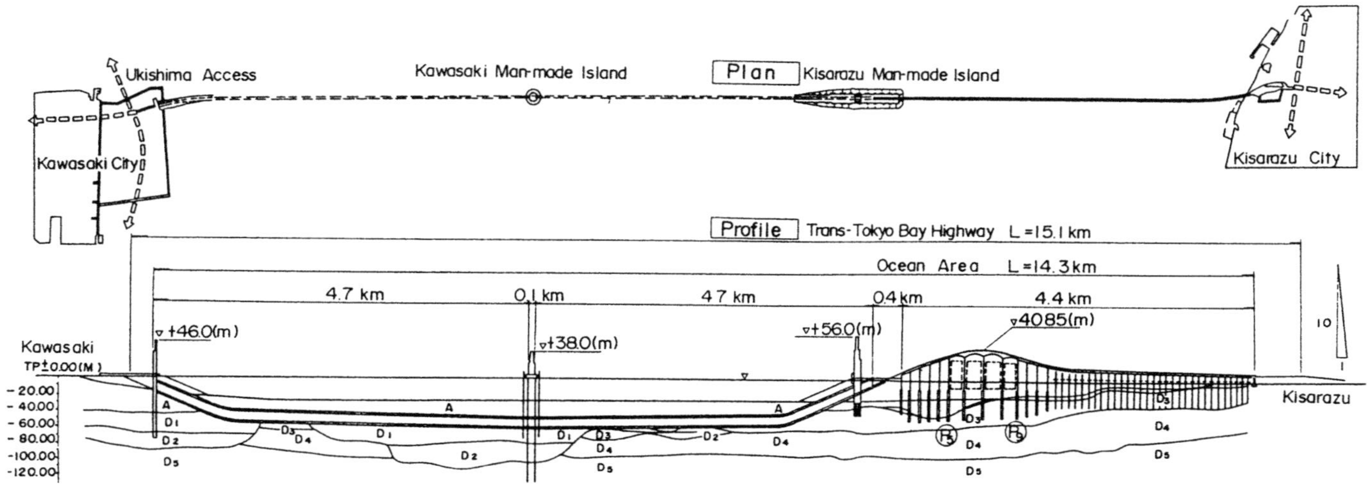


Fig.1 Trans-Tokyo Bay Highway

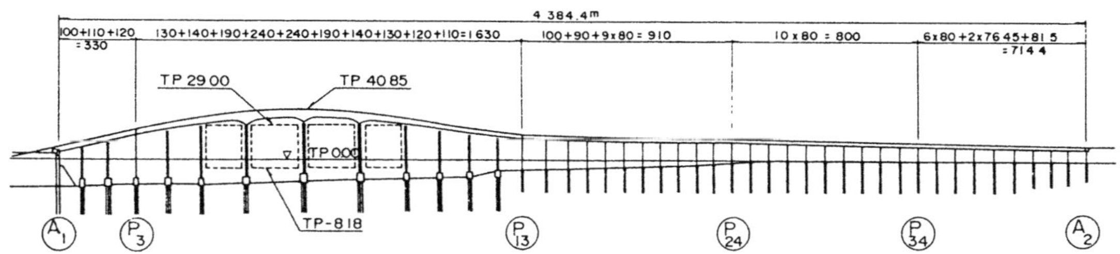


Fig.2 Layout of the TTB Bridge

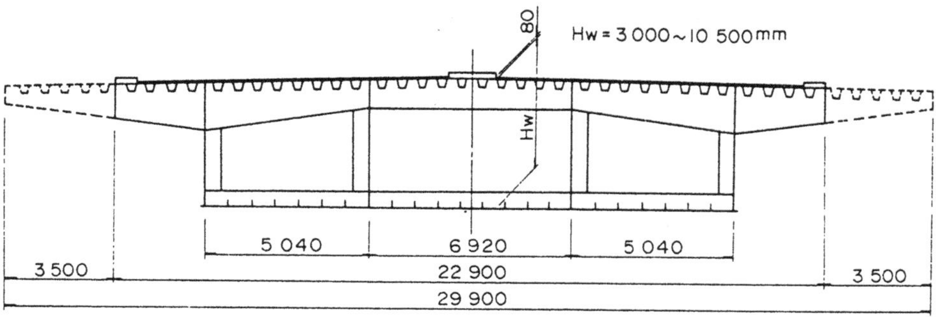


Fig.3 Steel Girder

highway alignment is considerably stiff with N-value over 50 and earthquake shear wave velocity over 400m/s, and has been taken as the input base of seismic design wave.

## 2. Structural forms of the bridge

The steel box girders with orthotropic decks have been adopted as superstructure because of its light dead weight and rapid construction. The maximum web height is 10.5m, and the deck width is 22.9m which accommodate dual two lane carriageway in the first stage, and will be extended to allow the extra two lanes in the future.

Composite steel piers with steel piles shown in fig.4 are constructed for the piers P1 to P12 in the 15m to 25m deep water area and will suffer strong horizontal forces due to ocean waves and earthquakes. Steel pile well foundation and concrete column with 6m by 4m cross section shown in fig.5 was selected for the piers P13 to P42 in the shallow water areas.

The bridge is designed to have long continuous girders so as to increase the seismic resistance by sharing seismic horizontal forces with many piers, and to improve road surface smoothness by eliminating expansion joints.

There are mainly three means to achieve continuous girders as shown in fig.6 (a) to (d). In system (a) only one pier is fixed to the girder, and therefore horizontal forces concentrate on it. When piers are flexible, system (b) is reasonable and more than one piers are fixed and horizontal forces can be distributed. In system (d) dampers do not resist slow motions due to temperature changes, whereas show large resistance to quick motions due to earthquakes. System (c) utilizes shear elongations of rubber bearings to absorb lateral girder motions. For the girders in deep waters where the piers are high and flexible system (b) is adopted, and system (c) is chosen for those in shallow waters where the piers are relatively low and rigid.

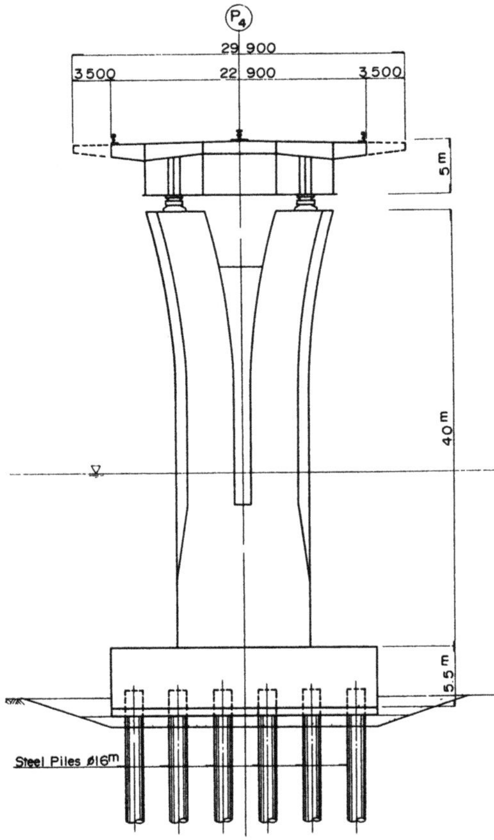


Fig.4 Steel Composite Pier

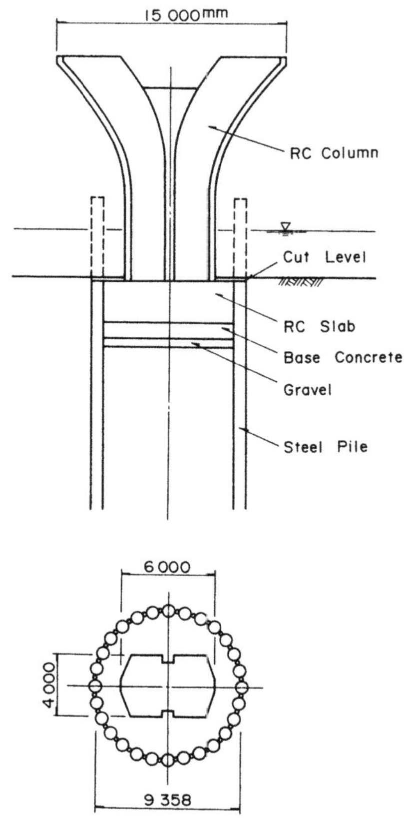


Fig.5 Concrete Pier

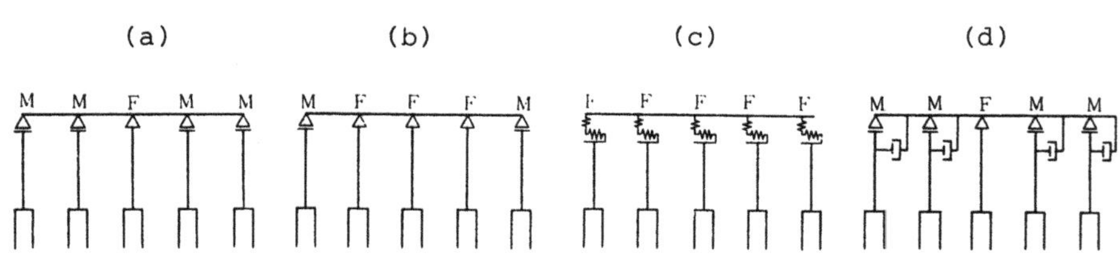


Fig.6 Measures to Achieve Multi-Continuous Bridges

3. Continuous bridge with flexible piers

For the bridge P3 to P13 the girder and the piers are fixed with hinge joints except end piers. Fig.7 shows bending moment, axial compressive forces and shear forces of girders due to dead loads, live loads and temperature change of 35°C. The bending moment caused by the temperature change is only 5% of the dead and live loads. The maximum compressive axial force at the center of the P7–P8 due to the temperature change is about 5,500tf which is equivalent to 10% of the total stress. The maximum girder vertical deflection due to live loads is 407mm, which is within the allowable value. The maximum vertical reaction of bearings occurs at P7 with 3,890tf, and no lift force exists. These calculation results above show the validity in achieving continuous girders.

Fig.8 shows the dimension of columns and piles of P4 to P12. The number of piles required for the middle piers such as P7 depends on the horizontal forces due to earthquakes, whereas those for end piers such as P3 and P10 depend on the lateral force caused by temperature change.

4. Continuous bridge with rubber bearings

For the bridge P13 to A2 in shallow water area piers are concrete and relatively stiff, therefore rubber bearings are selected to achieve multi-span continuous bridge. Three types of rubber bearings shown in fig.9 were thought to be feasible to absorb the large vertical and horizontal forces as well as long displacements: namely (a) lead rubber bearing (LRB), (b) high damping rubber bearing (HDRB), (c) ring rubber bearing(RRB).

Dynamic analysis was carried out to investigate the earthquake response to the bridge. Fig.10 is a simplified model and fig.11 is the design acceleration waves adopted for the TTB structures. This design wave is input from the base stratum and amplified through the soil layers. One of the interesting calculation results is shown in fig.12 . The peak response is 0.48Hz at the girder center, and is equivalent to the shear deformation of rubber bearings, therefore only rubber bearings deform laterally with piers at the stable position. The peak frequency of the pier top is 2.59Hz which is the eigen value of the pier, little effect of the superstructure was observed, and only the piers deform. In consequence the superstructure and substructures behave rather independently, and the bending moment at the pier bottom is less than static bending moment.

A couple of experiments were carried out to investigate the basic behaviors and strength of rubber bearings due to large lateral forces. Fig.13 shows the horizontal force and displacement relation of the high damping rubber bearing (HDRB). Histerisis was observed during cyclic motions. Picture 1 shows the deformation of the bearing with the lateral/vertical displacement ratio r of 220%. No damage is occurred during these tests.

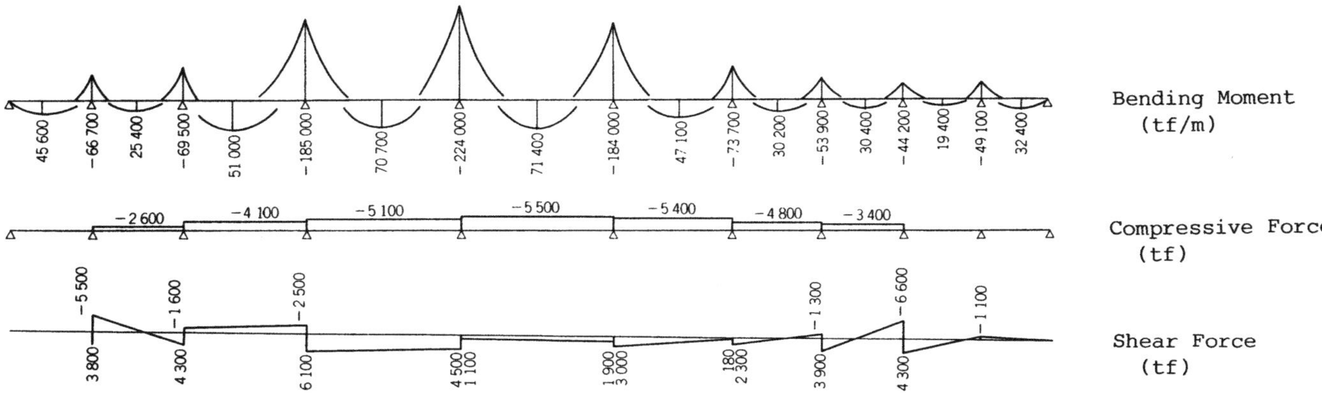


Fig.7 Sectional Forces of P3-P10

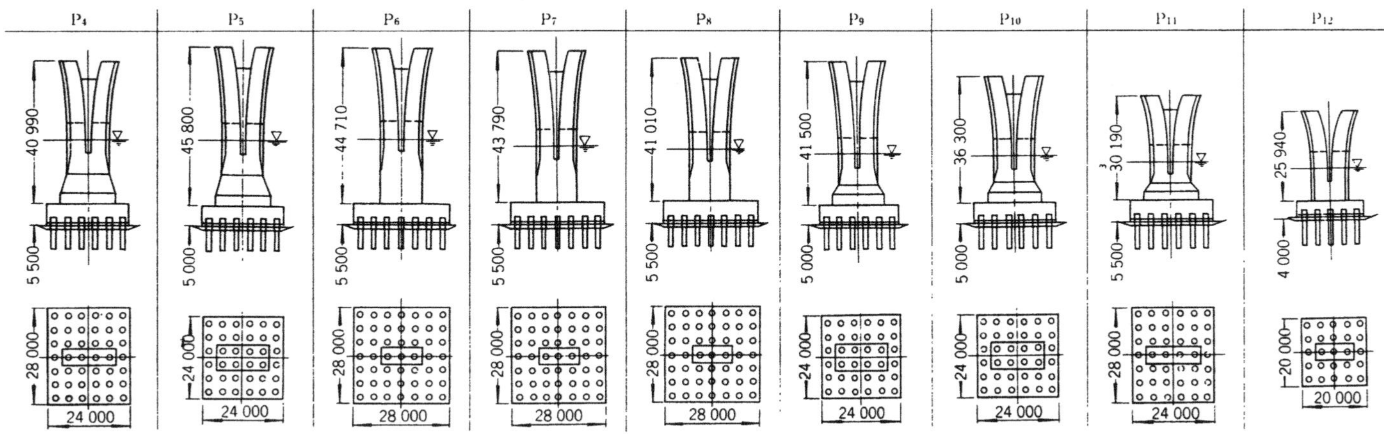


Fig.8 P4-P12 Piers (unit:mm)

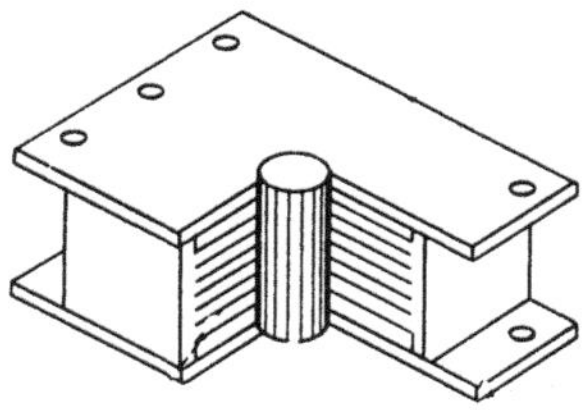

(a) Lead Rubber Bearing

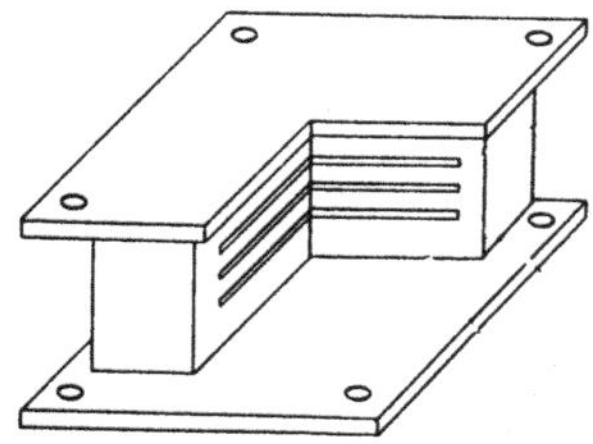

(b) High Damping Rubber Bearing

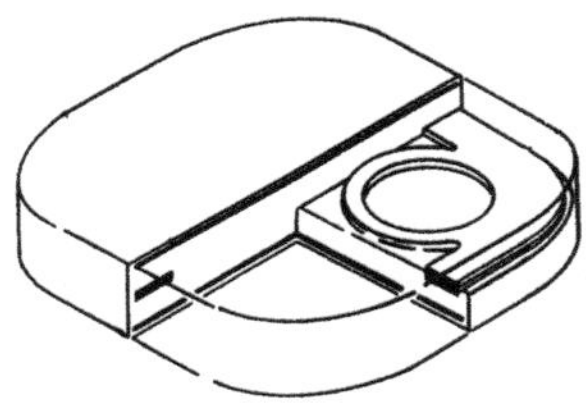

(c) Ring Rubber Bearing

Fig.9 Rubber Bearings

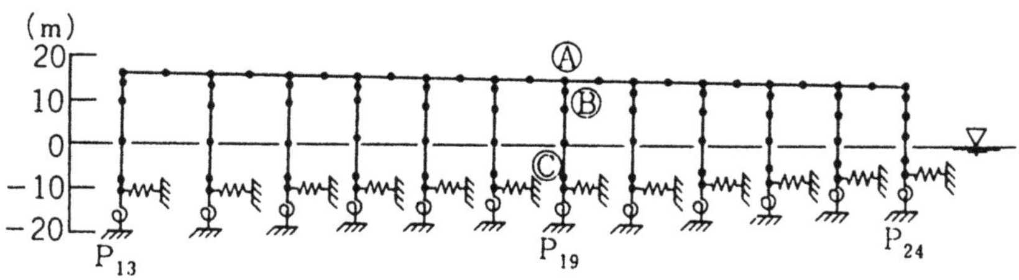


Fig.10 Dynamic Model
(A:girder, B:pier top, C:pier bottom)

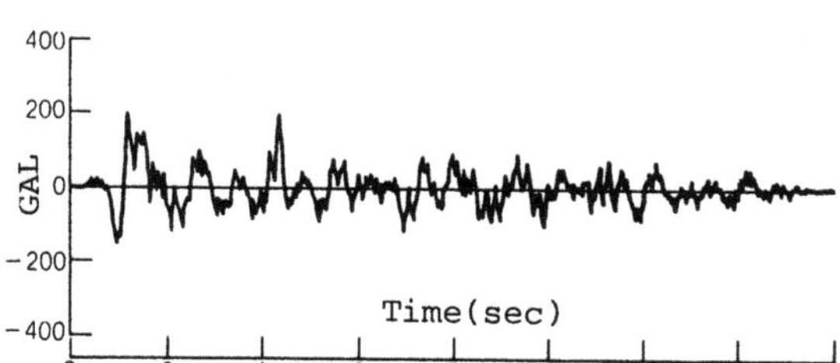


Fig.11 Design Seismic wave

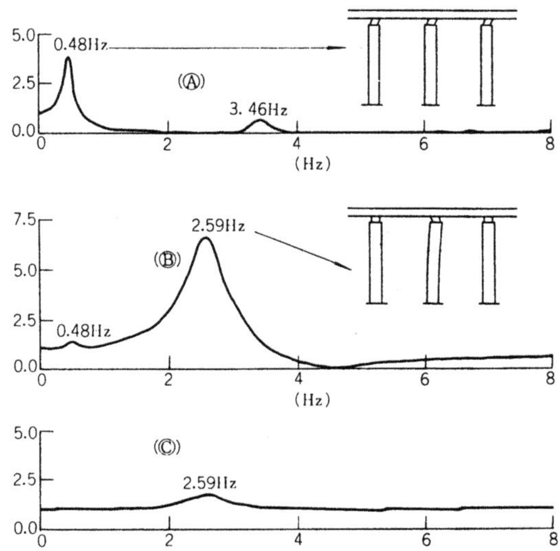


Fig.12 Frequency Response function

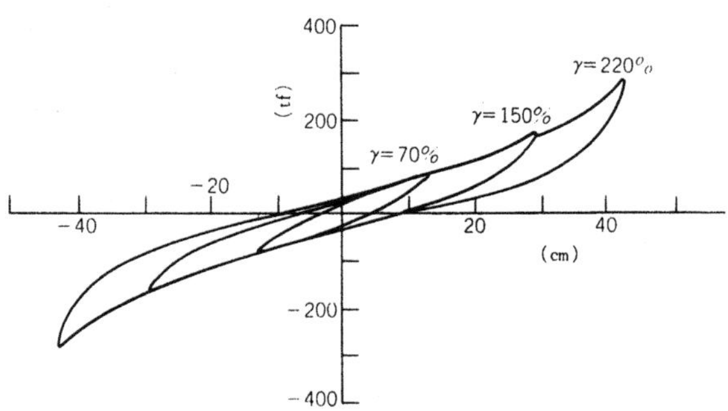


Fig.13 Lateral Behaviour of HDRB

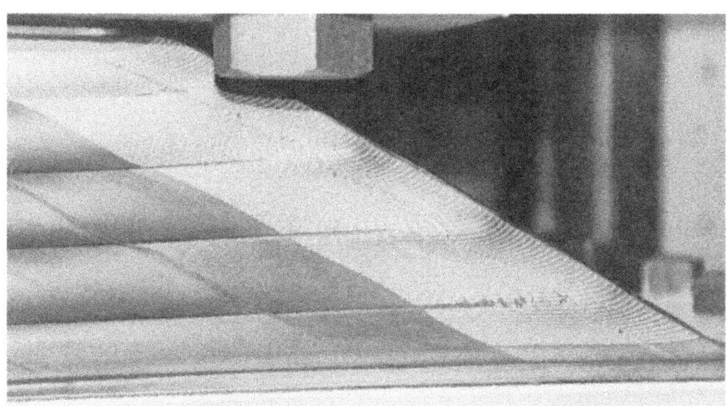

Picture 1 Lateral Deformation of HDRB

On one of the specimen surface a 10cm cut was made purposely, and the specimen was loaded vertically and laterally with r of 220%, but the cut did not develop.

## 5. Expansion joints

Table 1 shows the longitudinal movements of girder at P3 and P13 due to earthquake and temperature change. The design movement which expansion joint can accommodate is decided to absorb the 110% of calculated displacements plus 30mm to include additional movements which may result from constructional errors, girder rotation and minimum gap, and reaches more than 1m.

Table 1 Longitudinal movement of expansion joints

| | P3 | | P12 | |
|---|---|---|---|---|
| | (P2) | (P3) | (P11) | (P13) |
| EARTHQUAKE | 240 | 139 | 158 | 272 |
| TEMPERATURE (35deg) | 82 | 377 | 312 | 197 |
| EXPANSION JOINTS | 1030 mm | | 1155 mm | |

The cantilever comb joint shown in fig.14 is favored to accommodate such a long longitudinal displacements. The joint is made of cast iron, which has a good anti-corrosion property, and is bolted to the upper flange of the girder. As the structural form of this joint is simple, the replacement is relatively easy and noise complaint due to vehicles may not exist.

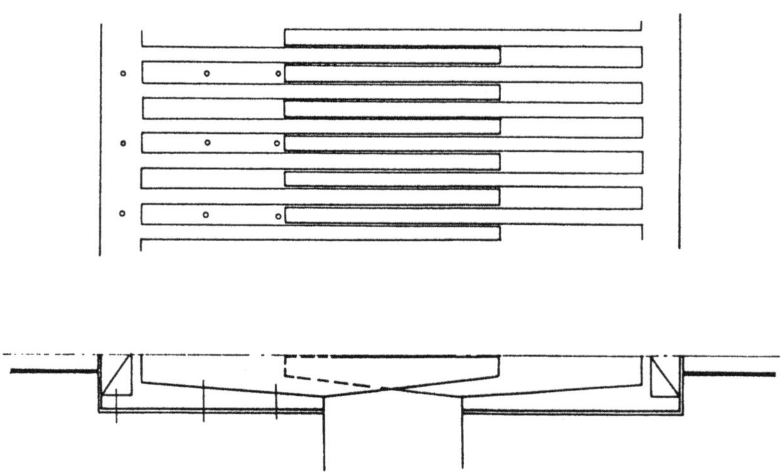

Fig.14 Comb Joint

The device to prevent falling the girder from piers in case of earthquake is attached at the girder ends to connect the both girders as shown in fig.15. This device is designed to work after the collapse of expansion joints and bearing stoppers.

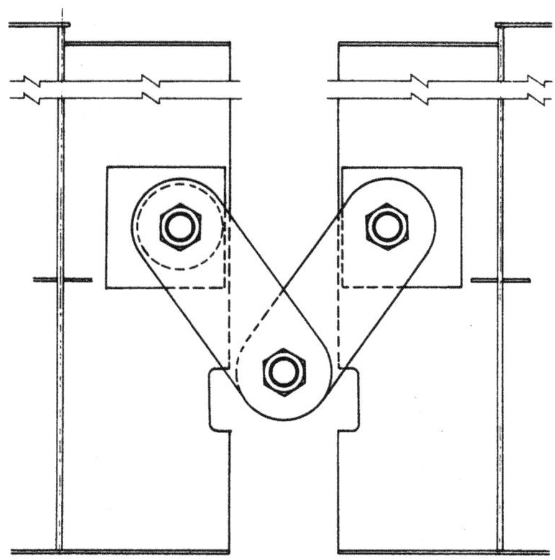

Fig.15 Bridge Falling Prevention Device

## 6. Conclusion

The structural forms of the very long continuous girder of the TTB Bridge is introduced in this paper. At present 6 piers have been already completed and the rest piers are now under construction. The superstructures are now being fabricated at factories. The whole highway will be finished and open by 1996.

The authors greatly appreciate Prof. M. Ito of Saitama University and Mr. H. Ikeda of Japan Highway Corporation for giving them the opportunity to participate in this colloquium.

# 9 CONTINUOUS STEEL AND COMPOSITE BRIDGES

J.D. PARSONS
Cass Hayward & Partners, Chepstow, UK

Abstract
This paper describes how the problems associated with expansion joints and bridge bearings in simply supported spans of multi-span steel structures can be overcome by the preferred and significantly more durable continuous steel and composite bridge construction. Totally jointless steel and composite bridges are discussed and integral abutments illustrated. Achieveable continuty, economy and aesthetics are also considered. Retrofit of existing bridge structures is mentioned.
Keywords: Continuous Steel and Composite Bridges, Durable, Economic, Aesthetic.

## 1 Introduction

Designers of steel and composite bridges have increasingly been moving towards joint-free continuous bridges for many years. In adddition to the reduction in maintenance liability associated with moving parts, i.e. expansion joints and bearings, continuous steel and composite bridges have other advantages such as economy both for the super and substructure and improved aesthetics.

Having recognised the problems associated with joints and bearings designers are giving attention to the ends of bridges, both single and multi-span, where inspection and maintenance galleries are provided often complicating construction of the abutments. The way forward appears to include the consideration of integral abutments connected to the deck which deletes joints and bearings altogether, removes the need for galleries and simplifies once again construction of the abutments.

Rehabilitation of existing steel multi-span structures is undertaken in other countries where a no-joint or minimal-joint policy already exists. To date there is little experience of such retrofit in U.K. but as strengthening of the bridge stock continues this may be the way forward for some bridges.

## 2 Continuity in the superstructure

### 2.1 The problem

The problems associated with expansion joints and bearings on simple

*Continuous and Integral Bridges*. Edited by B.P. Pritchard. Published in 1994
by Taylor & Francis, 2 Park Square, Milton Park, Abingdon, Oxon, OX14 4RN. ISBN: 0 419 19030 9.

bridges, including those of steel or composite construction, either single span or multi-span are well known. Maintenance and repair expenses have risen sharply in recent years and not insignificant sums have been spent on these areas.

Damage can be partially attributed to the increase in traffic volume, especially of heavy vehicles. Impact forces caused by vehicles passing over the expansion joints, is alone often sufficient to lead to problems. Unfortunately impact is often combined with poor detailing. Eventually water and de-icing salts penetrate through joints and leaks down onto the superstructure and substructure below. Damage occurs to the expansion joint and to the reinforced concrete slab ends. Ends of girders and steel bearings corrode and deterioration leading to unsightly paintwork and staining of the substructure. Reinforced concrete crossheads and piers also corrode following attack by chloride penetration.

Access for inspection to these vulnerable parts is usually difficult and repairs, which nearly always have to be performed while the bridge is in use, and maintenance, already restricted in many ways, tend to be further complicated.

A feature of simple multi-span bridges is the poor ride quality of the numerous and other frequent expansion joints (a notable example is the Midland Links).

In BS.5400:Part 10 Code of Practice for fatigue, allowance for impact is to be made. Where a discontinuity occurs in the road surface, e.g. at an expansion joint, the static stress at every point affected by a wheel, at or within 5m of the discontinuity, should be increased by magnifying the relevant influence line by 25% at the discontinuity reducing to zero allowance at 5m.

## 2.2 The solution

Since the 1970's most multiple span steel bridges have been made continuous both for dead loads and live loads over intermediate supports. This reversed the practice up to 1970 reported by Holland; of 28 viaducts put out to contract between 1967 and 1969 only 141 (15.1%) of 933 spans were continuous form for multi-span structures. Experience is that continuous bridges decks are significantly more durable then those consisting of simply supported spans. The Department of Transport strongly recommends the use of continuous decks wherever possible.

## 2.3 The practice

Spans are usually fixed by site restrictions or obstacles and required clearances. Where freedom exists an economical span in the range 25m to 45m is likely. However, where deep piled or other expensive foundations are needed, budget costing will encourage the use of longer spans.

Multiple spans of approximately 26m suit universal beams. The optimum for using plate girders for continuous bridges is about 45m, because 27m long "span girders" which should be a lighter section, can be spliced with "pier girders" of a single plate 18m long. For longer spans, more shop or site splices are needed. Component lengths for shop fabrication should be the maximum possible consistent with delivery and site restrictions to reduce the amount of on-site assembly. The maximum length for road delivery without restrictions in the U.K., is normally 27.4m although longer lengths can be transported by arrangement.

Generally the girders run over the piers and are site spliced within the span near the points of contraflexure. Site splices may be formed with high strength friction grip bolted joints. Site welded joints are sometimes used by erection is less rapid.

Ideally the length of end spans should be about 0.8 of the penultimate span. However, because of site conditions continuous bridges are often curved in plan, have irregular span, or skews. With these structures there can be a tendency for negative reactions which can further exacerbate the damage at bearings or supports due to impact, vibrations or reversal of stresses.

In continuous bridges with integral steel crossheads the concrete deck slab at intermediate supports is likely to be more heavily reinforced due to the combination of longitudinal global and local bending. The control of crack widths here, important for long term durability, may govern the reinforcement design.

## 3 Continuity with the substructure

### 3.1 Integral steel crossheads

Earlier composite steel bridges tended to use closely spaced multiple girders (1.5m to 3.5m spacing) with leaf piers supporting every girder on many bearings. For wide decks such piers are wasteful and aesthetically undesirable so that from the 1970's column type piers

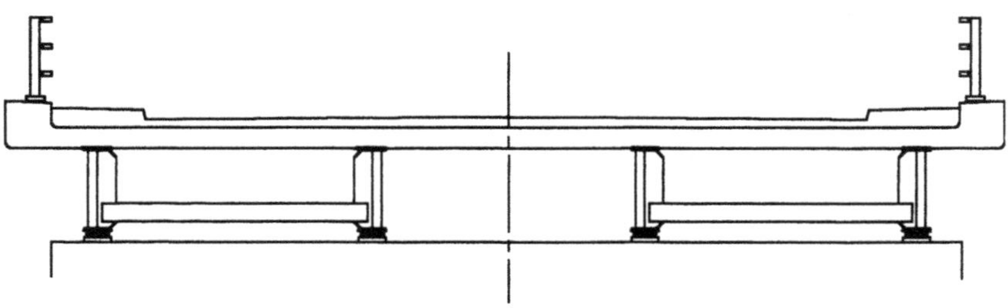

Fig. 1a Multiple UB deck on leaf pier

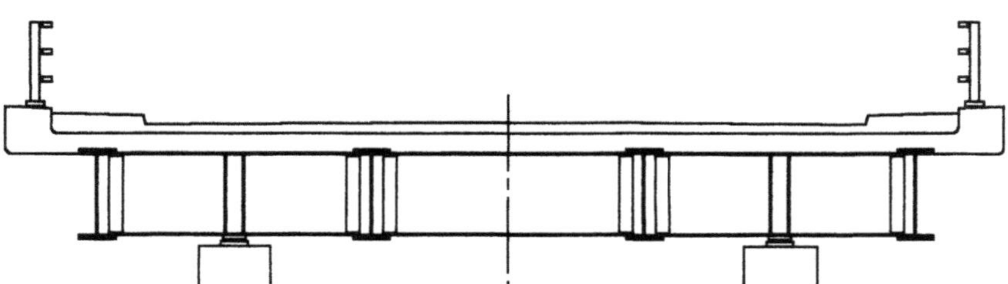

Fig. 1b Multiple UB deck with integral steel crosshead

surmounted by steel crossheads integral with the deck have often been used. Fewer bearings are demanded which with a crosshead shallower then the main girder can be raised into a more sheltered environment (see fig.1).

Examples are the M27 River Hamble Bridge, M25 Mar Dyke Viaducts, Skipton Bypass, M25 Brook Street Viaduct and M25/A1(M) South Mimms Interchange Bridges. A recent example is the Blaydon Haughs Viaduct (A1 Newcastle Western Bypass) which won a Steel Award in 1991.

### 3.2 Flexible piers

When intermediate supports are tall or can be slender then moving bearings can be avoided altogether. The movements caused by thermal expansion and contraction are accommodated by bending of the piers. Other design load effects such as longitudinal braking or traction can be shared between the piers in proportion to the pier stiffness.

Examples are the M5 Avonmouth Bridge where flexible piers 1.2m x 4.27m range in height from 13m to 33m and bearings are the fabricated knuckle type, and the recently constructed pair of 3 span River Avon Bridges (Bristol Spine Road) where fixed bearings are employed at each of the intermediate piers (1.7m x 1.0m and 7m high).

### 3.3 Integral abutments

Completely jointless steel bridges, i.e. continuous span bridges without expansion joints at either end due to the connection of the superstructure to the abutment, are rare in the U.K. but have been adopted for a number of bridges in the U.S.A. Wasserman (1986) reported that jointless bridges had been constructed for the past 20 years and that 80% of bridges in Tennessee were being constructed without expansion joints. This included composite steel bridges of up to 127m length. It was admitted that the practice seemed to defy the calculations which showed that local damage or cracking should have occurred at abutments.

In attempting to justify the use of jointless bridges it should not be forgotten that many older U.K. structures, especially railway bridges, had no expansion facilities at all, although if multiple span, they were usually simply supported so that some flexibility was achieved between one span and another, even if unintentionally. Girder ends were usually built in at abutments without bearings or joints. Such structures have often experienced no problems and appear to defy calculation, although most are of short length compared with modern viaducts.

From the U.S. experience it would appear to be feasible to use jointless steel deck in the U.K., where the temperature range is generally less severe. However, conditions may not always be suitable, as when abutments are comparatively rigid. The concept envisaged by Hambly and Nicholson (1990) whereby the abutment slides on supporting granular fill implies on supporting granular fill implies movement such that a roadway expansion joint may still be desirable or regular repair of the surfacing locally will be necessary. Figure 2 shows possible abutment details for jointless steel bridges. Figure 2(a) requires the abutment to slide on granular fill, Figure 2(b) involves flexure of piles and Figure 2(c) incorporates a joint, but which is buried at deck soffit level. The

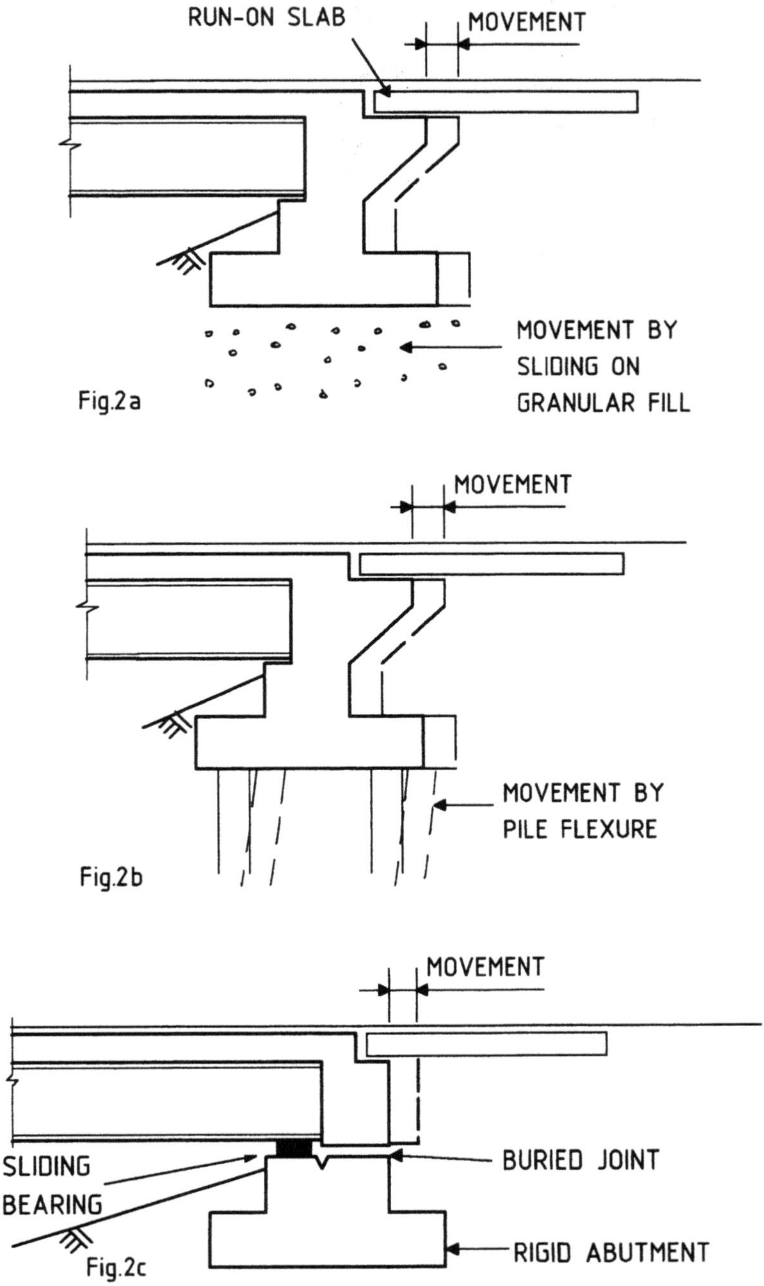


Fig.2. Abutment details for jointless steel bridges

arrangement of 2(c) has been used on railway bridges with no apparent difficulty. It implies that the fill behind the deck must compress to take up the movement. In all cases, the steel girder ends are cast into the abutment or curtain wall which is easily achieveable, but it is desirable to incorporate a weather flat at the steel/concrete interface to prevent undetected corrosion of the steel web.

An additional advantage of integral abutments is that the inspection gallery required for inspection maintenance of expansion joints and bearings can be removed simplifying once again the construction of the abutment. However, reinforced concrete run-on slabs are required to prevent excessive compaction of the fill behind the abutment.

## 4 Poor ground and continuity

Traditionally for mining areas or whenever the interpretive soils report warned of a risk of relative settlements, simple spans were used. In the computer we how have the tool to make designing continuous bridges in such circumstances manageable and the designer can make due allowance in the design for possible displacements in the dead load moments and shear effects due to settlements. The use of slender decks reduces the effects of settlements.

## 5 Aesthetics

A continuous steel or composite bridge gives the designer scope for producing an elegant structure of clean appearance with pleasing lines. Welded site splices can improve appearance for bridges in sensitive areas. Pairs of stiffeners at supports can be 'boxed' in. Integral steel crossheads can be used to reduce the number of piers and raise the bearing out of sight.

Curved girder soffits also greatly improve the line and interest of the structure as well as making economic use of often increased construction depth adjacent to piers. Fig. 3 shows the aesthetic benefits of a continuous girder bridge compared to a simple girder structure (as can be seen at the Midland Links) as well as reducing the number of expansion joints and bearings and reducing size of sub-structure.

Curved bridges are dealt with by making the girders from straight chords between site splices, but an increasing U.K. trend for sharper radii is to use curved girders taking advantage of modern fabrication methods.

Examples are the Simon de Montfort Bridge, Evesham Bypass (Award 1988) and Festival Park Flyover, Stoke on Trent (Award 1990). Incidentally, both examples are of twin plate girder and cross girder type form with only two bearings with supporting columns at intermediate piers.

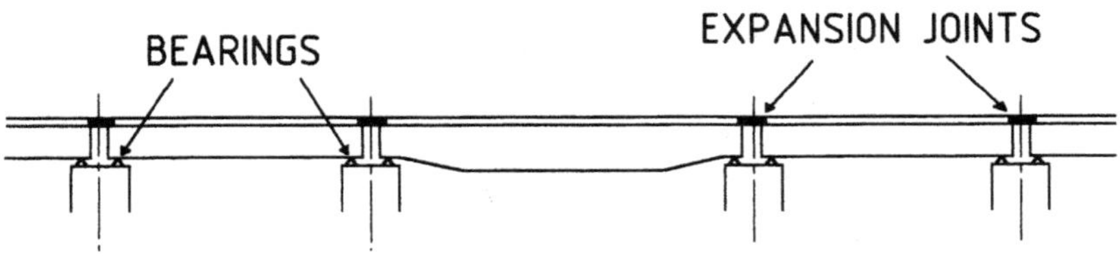


Fig.3.a. Simple girder structure

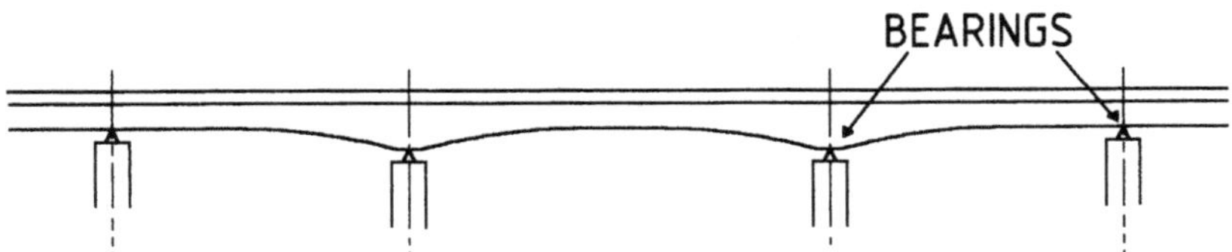


Fig.3.b. Reduction of number of bearings and expansion joints by use of aesthetically pleasing continuous curved soffit girders

Fig.3. Continuous girder structure example

## 6 Erection

In addition to crane erection of full span lengths continuous bridges can be successfully erected by the launch method. Although this method requires the girders to have a level soffit site assembly of the individual girder elements are at ground level and craneage requirements are significantly reduced.

Examples are Loughor bridge, Llanelli and Chepstow Viaduct (1987) with a main span of 93m and overall length 302.5m.

## 7 Retrofit continuity

The current programme of bridge strengthening and repairs gives scope for reducing or improving future maintenance on existing structures. A suitable rehabilitation method may be based on one used in Australi for new designs, where continuity for live loads only is achieved for simply supported steel girders which are concreted in over the piers with rebars in the deck slab forming tensile continuity. This could work well where deteriorated deck slabs necessitate concrete removal.

Experience in Tennessee, U.S.A. is that for short span steel bridges no girder end modification is required. For longer steel pan compression flanges are jointed. In some cases, bearings must be modified. Where the foregoing modifications are employed, if structure widening is to be added, the new portion will be of fully continuous composite construction.

## 8 Conclusion

Multi-span steel bridge should always be made continuous because this is easier to achieve and is economic. The use of integral abutments to achieve a completely jointless deck for both single and multi-span bridges should be considered further taking into account the success of the practice in other countries. The economy of repairing the road surfacing compared with provision of joints requires consideration.

The amount of theoretical movement which can be tolerated without joints is tentatively suggested at ±25mm for surfaced carriageways and ±40mm for ballasted rail tracks. For structures fixed at one end this implies a maximum bridge length of about 70m for highway bridges and 120m for rail bridges. These lengths can be doubled if the structure is fixed at mid length.

Continuous steel and composite bridges can be both aesthetically pleasing and durable for the future. Aesthetis is important, but maintenance is of greater concern.

## 9 References

Hambly, E.C. and Nicholson, B.A. (1990) **Prestressed beam integral bridges.** The Structural Engineer, December

Holland, A.D. and Deuce T.L.G. (1970) **A Review of Small Span Highway Bridge Design and Standardisation.** Journal of the Institution of Highway Engineers, August

Wasserman E.P. (1986) **Jointless Steel Bridges.** National Engineering Conference, AISC, Nashville

# REPORT ON SESSION THREE

G.P. TILLY

Dr Nakamura explained that the trans Tokyo Bay bridge was designed in continuous sections, each composed of 9 to 11 spans. The crossing includes a suspension bridge having a span of 1990 m.

In response to a question about how the rubber bearings described in the paper have been designed to have higher values of damping, the author explained that constituents had been added to the rubber when it was manufactured, so that the resulting material of the type (b) bearing shown in Figure 9 had increased the hysteresis. The type (c) bearing had a steel ring which strengthened the connection between the rubber and steel plate. This bearing did not have high damping.

It was commented that experience of the tuned dynamic absorbers that are sometimes used to dampen lively bridges was that they are sensitive to temperature and there can be problems with durability. In general, design lives of bridges can be as high as 100 years but it should be recognised that joints and bearings cannot be expected to perform satisfactorily for as long as this and designs should take into account the need to fit replacements at suitable intervals.

In response to a question about composite construction, Dr Nakamura said that in Japanese bridge design, composite steel girders are not used because designers are not fully satisfied that they have enough information to cope with hogging moments. A speaker commented that in the United States this is not regarded as a problem. Designs of composite and non-composite bridges are similar and no account is taken of creep and shrinkage. In the past it had been the usual practice to cast the in situ concrete in a correct sequence in order to balance the loads but the concrete is now machine laid and is placed in a continuous run, from one end of the bridge to the other, without stopping. Transverse cracks are sometimes developed in the concrete but this is not regarded as a problem.

Dr Parsons presented his paper on continuous steel and composite bridges. He discussed the movement of integral abutments and suggested that it occurs by a sliding mechanism. During the discussion a speaker questioned whether rotation is more likely to occur. The author felt that the movement was related to stiffness of the abutment and there could be creep of the soil. This type of design is known to perform satisfactorily in practice but he was able to provide an adequate explanation of the mechanics.

A speaker commented that in France bank seats are used with a single line of piles to reduce movement capacity and permit rotation. There are 150 such bridges on the Alpine roads and they have now been performing satisfactorily for 12 years. In France run-on slabs are commonly used but are designed with a joint between the bridge deck and the slab. The running surface is laid continuously over the joint to provide a smooth ride.

In the UK there were problems with run-on slabs breaking due to settlement and their use was abandoned about 10 years ago. It is now required that they be designed to be cantilevered or supported on piles. In the United States 7 to 9 m long run-off slabs are commonly used. The soil beneath is compacted according to normal requirements and it is not considered necessary to specify the use of granular fill. The different States follow their own practice.

# PART FOUR

# INTEGRAL BRIDGES

# 10 DANISH INTEGRAL BRIDGES

E.M. VEJE
Cowiconsult, Consulting Engineers and Planners AS, Lyngby, Denmark

Abstract
This paper gives a description of the recent major Danish bridge projects, especially the Great Belt Link Project, focusing on design developments from the Little Belt Suspension Bridge, opened for traffic in 1970, to Great Belt East Bridge, 'set for construction' completion 1997/98 with special attention to the efforts made to maximize the length of the continuous superstructure sections.

Keywords: Farø Bridges, Great Belt Link, Expansion Joints, Dehumidification system, Closed Box Girders

## 1 Introduction

In recent years designs of multi-span bridge superstructures show a clear trend towards long, continuous sections. This is in acknowledgement of the benefits of high driving comfort and reduced maintenance effort when the number of expansion joints in the structure and the roadway pavement is limited.

This design trend is driving the development of expansion joint structures and bridge bearings towards increased movement capacity. Vice versa, the capacity of these special components sets limits to the possible continuity length of the structures.

## 2 The Farø Bridges

The Farø Bridges comprise two bridges, a northern bridge from Zealand to Farø and a southern bridge from Farø to Falster. The northern bridge comprises 18 spans of 80 m length, plus two end spans of 78 m. The southern bridge crosses the navigation channel by a cable stayed bridge situated approximately midway. It has a central span of 290 m for ship passage and side spans of 120 m each. At the ends of the cable stayed bridge 15 approach spans with the selected standard length of 80 m (except for the two end spans) complete the bridge (7 and 8 spans respectively). The bridge was opened for traffic in 1983.

The superstructure is outlined as a closed steel box girder, with almost the same outer shape on the entire bridge length. The closed steel box girder concept, which also was applied on the Little Belt

*Continuous and Integral Bridges*. Edited by B.P. Pritchard. Published in 1994 by Taylor & Francis, 2 Park Square, Milton Park, Abingdon, Oxon, OX14 4RN. ISBN: 0 419 19030 9.

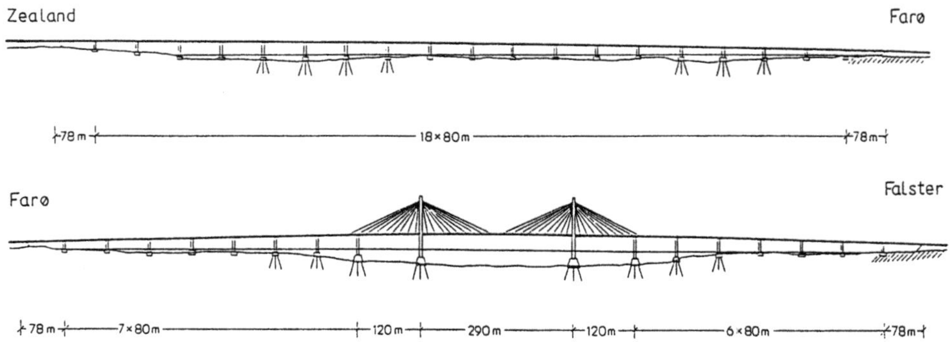


**Figure 1** : The Zealand-Farø and Farø-Falster Bridges

Suspension Bridge, built 15 years earlier, contains several features fundamental for cost efficiency :

- great uniformity of the individual elements of the box girder, thereby facilitating industrialized fabrication
- single cell box girder cross section without traditional transverse and longitudinal girders
- simple connection details between transverse bulkheads (per 4 m) and the exterior girder panels, thereby reducing labour consuming assembly tolerance problems
- corrosion protection by means of dehumidification of the interior and a smooth, minimized outer surface area to be painted; a concept successfully introduced on the Little Belt Suspension Bridge

Another characteristic for the Farø bridges is large continuous jointless length of the bridges - probably larger than for any other bridge at the time of construction. The two bridges are continuous from abut

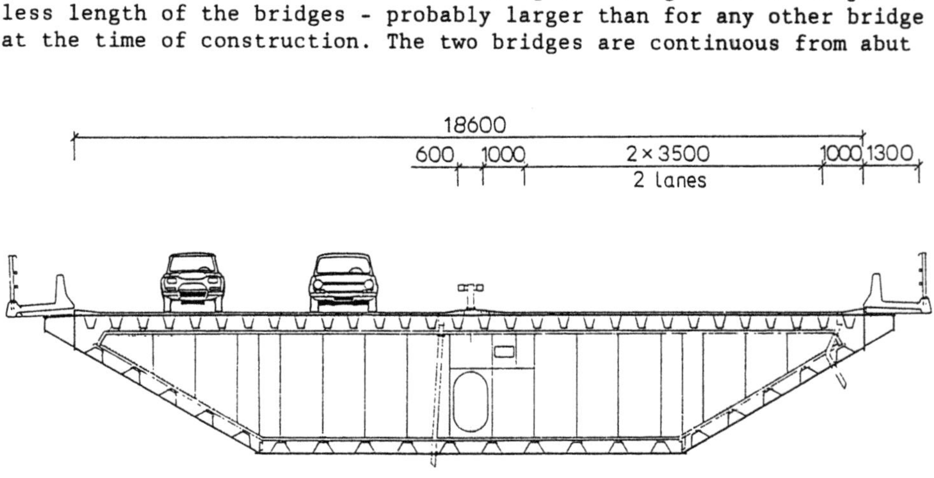


**Figure 2** : Closed steel box girder

ment to abutment, corresponding to expansion section lengths of 1596 m and 1726 m respectively, and have only expansion joints at the abutments.

A continuous structure is favourable in several respects:

- many service and maintenance problems are eliminated,
- traffic comfort is improved remarkably,
- construction costs are reduced.

The large expansion sections imply large movements, which also increase the wear of the bearings. For that reason the accumulated travel at each abutment is measured by means of displacement transducers. This gives useful information about the wear of the bearings and is of importance for the determination of the inspection intervals.

The extreme movements at the bridge ends amount to up to 1050 mm per joint, primarily caused by temperature variations.

With reference to the proper functioning of the bridge the choice of appropriate bearing types becomes a matter of great importance under such extreme conditions. The bearings must be reliable, solid and durable and their resistance to movements must be small, if a problem-free functioning is to be attained throughout a long span of time.

The installation of the bearings shall be placed very accurate, hereunder the alignment of the bearings and side guides.

The fixed bearings absorb the longitudinal forces due to braking and wind loads. Most important for the girder are the longitudinal forces that might result from unequal friction in the bearings of the two bridge halves. Experiences gained so far regarding the accumulation of frictional forces over a great number of bearings may be useful for the possibility of allowing longer bridge sections to be built without joints. The observations are envisaged to be continued over a period of 20-25 years, with a status every 5th year, so that an adequate basis for evaluation may be established by experience.

Special springs have been arranged at the fixed bearing allowing for measuring the horizontal force acting, primarily originated from active friction and friction differences. Measurements so far performed have revealed longitudinal forces in the order of 30% of the expected forces and amount to less than 10% of the maximum force that can be restrained. In the design it has been assumed that the difference in resistance between the bearings may be 2% of the vertical load.

## 3 The Great Belt Link

The 18 km wide Great Belt is divided into two main channels, east and west, by the tiny island of Sprogø, Therefore, it was obvious to take advantage of this island as an integrated part of a fixed link project, a 'stepping stone' between east and west across the water. The link consist of three major structures : A bored railway tunnel and a high level motorway bridge across the eastern channel, and a low level dual mode bridge for railway and motorway across the western channel.

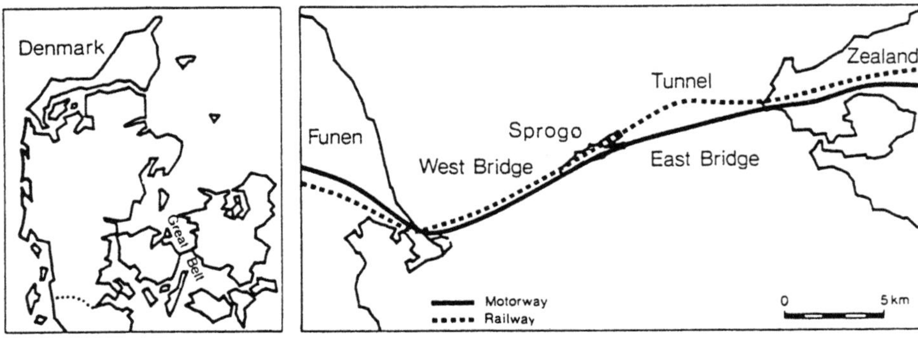


**Figure 3** : The Great Belt link

### 3.1 The West Bridge

The 6.6 km bridge consists of two haunched concrete box girders with variable depth each supported on separate pier shafts sharing a common substructure designed as a gravity founded caisson. The northern girder carries the railway track, and the southern the road traffic. The railway girder has an overall depth of 8.70 m, and 5.13 m at mid span. The roadway girder depths are 7.34 m and 3.78 m respectively.

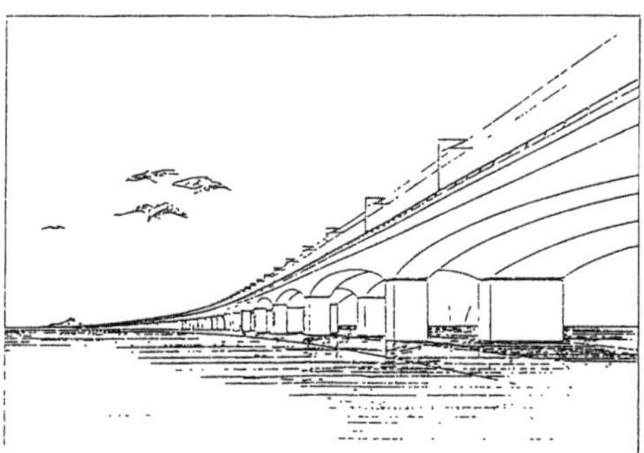

**Figure 4** : The West Bridge

The superstructure is divided into 51 main spans of 110.40 m and 12 expansion joint spans of 81.75 m. Expansion joints are provided at the abutments and at 4 intermediate piers, thus subdividing the overall length into six continuous girders of about 1100 m.

The railway expansion joint is an essential part with regard to the safe operation of the railway. The joint used on the West Bridge is a new development and it has been decided to install and test a prototype of the rail expansion joint in the Danish State Railways' main line.

The expansion joints connecting the continuous superstructure of about 1100 m length will move between ±350 mm and ± 600 mm. A total capacity of up to 1200 mm has so far only been seen in Japan for the Honshu-Shikoku Bridge project.

The requirements for angular rotation, vertical radius of the rail and comfort level are parallel requirements focusing on a high comfort level and high safety of operation. The railway expansion joint arrangements is capable of meeting these requirements, and based on developments of experience gained in Japan and Germany.

### 3.2 East Bridge, Approach Bridge

The East Bridge is designed with extremely large jointless sections, probably the longest ever performed in bridge design. The entire 6.8 km bridge is only subdivided into 3 expansion sections of 2.7 km, 2.5 km and 1.5 km.

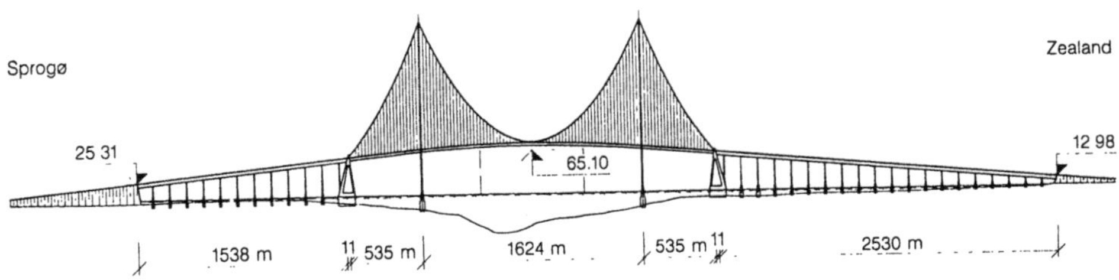


**Figure 5** : The East Bridge Elevation

The two approach bridges are 1538 m and 2530 m long respectively, with intermediate span lengths of 193 m and end spans of 140 m and 62 m. The approach bridge girder is based on closed box sections with a central longitudinal bulkhead which minimize the shear lag effect and ensures shear capacity in case of ship collision against one of the inclined sides. The interior of the girders (both approach and suspension bridge) is unpainted and protected by dehumidification as for the Farø bridges. About 80 % of the total steel surface is thus unpainted with substantial cost and time savings as a result. The depth of the girder is 6.7 m.

Each approach bridge is continuous from the abutment to the anchor block of the suspension bridge. The corresponding max total movement of the expansion joints is from 790 mm up to 1530 mm. Watertight modular expansion joints of the so-called Swivel Joint type have been selected as a result of intensive tests to achieve the required 100 year fatigue life. To support the relatively flexible steel superstructure on the piers, spherical bearings have been chosen. The advantage of these bearings is the relatively low restraining moments originating from rotation friction in the spherical parts, which is of importance for the design of the supporting structures in the bridge girder. The horizontal movements are transferred by sliding over a spherical segment. The sliding medium is PFTE (Teflon) against stainless steel.

### 3.3 East Bridge, Main Bridge

The navigation span of the East Bridge crosses the international shipping route between the Baltic Sea and the North Sea. Based on intensive studies of cable stayed and suspension bridge concepts as well as navigational risks a suspension bridge with a 1624 m main span and two side spans of each 535 m was decided upon.

The main span is outlined with a cable sag corresponding to 1/9 times the span length. Vertical hangers each 24 m support the girder. The suspension bridge girder is outlined as a closed box girder as for the approach bridge, with depth of 4 m and width of 31 m. The bridge is arranged with an innovative articulation around the pylons allowing the girder to be continuous over the full length of 2.7 km between the two anchor blocks. The traditional expansion joints at the pylons are avoided. Compared to a system with joints at the pylons, analyses have indicated an approximately 25% reduction in the longitudinal deflection of the girder from traffic load.

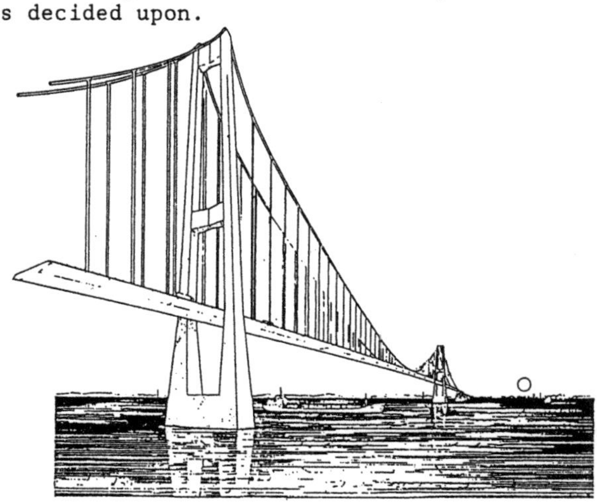

**Figure 6** : Architect's view of the East Bridge

If free movements were allowed, the extreme horizontal movement from the characteristic traffic load at the expansion joints would be 1.8 m. In order to limit longitudinal movements, hydraulic buffers are arranged between the anchor block and the girder. The buffers allow for slow horizontal movements, for instance due to temperature, up to ± 1.0 m and a free rotation of the girder, while fast movements, for instance due to a heavy vehicle passing the bridge, are restrained in order to increase the fatigue life and to reduce the wear on the expansion joint. The remaining portion of the theoretical movement will be balanced by restraining forces in the girder.

The advantages of the continuous girder concept are:

- A very simple arrangement at the pylons.
- Reduced installation and maintenance costs for the expansion joints at the anchor blocks
- Improved stiffness of the overall suspension bridge system.
- Improved aerodynamical stability.

The continuous girder concept also permits elimination of the traditional cross beams on the pylons immediately below the girder, and thus contributes to the aesthetical appearance of the bridge.

## 4 References

Haas, G. Petersen, A, Ostenfeld, K.H. : The Farø Bridges, New Developments in Design and Construction, International Symposium on Strait Crossings, Stavanger, Norway, October 1986. COWI-publication no 556.

Jacobsen, A.S. Petersen, A Sørensen, O : Design of the 3 km Continuous Steel Bridge for the Great Belt East Bridge, 10th Annual International Bridge Conference, Pittsburgh, Pensylvania, June 1993. COWI-publication no 869.

Ostenfeld, K.H. : Important Structures and their Specific Problems - Great Belt Link, Denmark, Seminar on Resent Developments in Bridge Design and Constructions, Athens, Greece, May 1991. COWI-publication no 756.

Ostenfeld, K.H. : Bridges of Today and Tomorrow, 10th Annual International Bridge Conference Pittsburgh, Pennsylvania, June 1993. COWI-publication no 868.

# 11 PRECAST SEGMENTAL CONSTRUCTION

J. MULLER
Jean Muller International, Montigny le Bretonneux, France

**Abstract**
This overview of precast concrete segmental construction includes: large urban highway or mass transit structures; a composite (steel and concrete) space truss segmentally built for medium span viaducts or very long cable-stayed spans; a special launching gantry application in long-span girder bridges and the monitoring of expansion joints in the first attempt at intelligent structures; three current concrete cable-stayed bridges; the new bi-stayed concept allowing clear spans of up to 2,400 m. The following paragraphs describe some salient features of current specific projects as an outline of the method's state of the art while projecting its potential development.
Keywords: Precast segmental construction; railway viaducts; highway bridges; composite construction; long-span girder bridges; cable-stayed bridges; space trusses; bi-stayed bridges.

*This paper is based on the paper given at the FIP Symposium, Budapest, May 1992*

## 1 Brief survey of past experience

Precast concrete segmental construction has now over 25 years of experience. After the precast arch bridges designed by Eugene Freyssinet and built by Campenon Bernard between 1946 and 1950, the world's first precast concrete bridge with match-cast joints, designed by the author, was built near Buffalo, NY, in the USA in 1952. Choisy-le-Roi bridge near Paris over the River Seine, designed and built in 1962-63 was then the first application of epoxy resin joints in match-cast construction and represented the actual start of precast segmental construction on an industrial scale.

Many bridge structures followed and were the occasion of successive innovations and developments. Significant steps are as follows:

1. Match-cast dry joints - New York, 1952
2. Epoxy match-cast joints - Choisy-le-Roi, 1962
3. Long bed precasting - Choisy-le-Roi, 1962

*Continuous and Integral Bridges*. Edited by B.P. Pritchard. Published in 1994 by Taylor & Francis, 2 Park Square, Milton Park, Abingdon, Oxon, OX14 4RN. ISBN: 0 419 19030 9.

4 Continuous deck built in cantilever - Choisy-le-Roi, 1962
5 Flexible-walled piers - Choisy-le-Roi, 1962; Juvisy, 1965
6 I-shaped piers and double-walled piers - St André de Cubzac, 1977; Magnan, 1973
7 Short bed precasting - Pierre Bénite, 1964; Paris Ring Road Bridges, 1965
8 Beam and winch placing - Pierre Bénite, 1964
9 Launching gantry - Oléron, 1964
10 Short-bed precasting with variable geometry - Chillon, 1965
11 Launching gantry with variable geometry - Chillon, 1965
12 Progressive placing and swivel crane - Rombas, 1972 Fontenoy, 1975
13 Multiple keys in match-cast joints - Alpine Motorway bridges, 1971
14 Assembly of full spans on falsework - Alpine Motorway overpasses, 1971
15 Concrete cable-stayed bridge with centre suspension - Brotonne, 1973
16 Precast segmental concrete urban railroad bridge - Marne-la-Vallée, 1975
17 Long-span segmental railroad bridge - Clichy, 1976
18 External prestressing concept studies, Long Key Bridge, Florida, USA, 1976 (Fig. 1)
19 Span-by-span construction on assembly truss; Transverse prestressed deck slab; as-cast riding surface - Long Key Bridge, 1976
20 Precast segmental piers - Linn Cove, NC, USA, 1977

## 2 Recent developments in medium span girder bridges

### 2.1 San Antonio Y Project, TX, USA

Currently, this highway viaduct represents the single largest application of precast segmental construction in a difficult urban environment. The use of a shallow, narrow box girder with large overhangs (see typical cross section in Fig. 2) contributes to the aesthetics of this unique project. External prestress and span-by-span construction with self-launching assembly trusses have been used to allow maximum construction efficiency and low cost, Table 1.

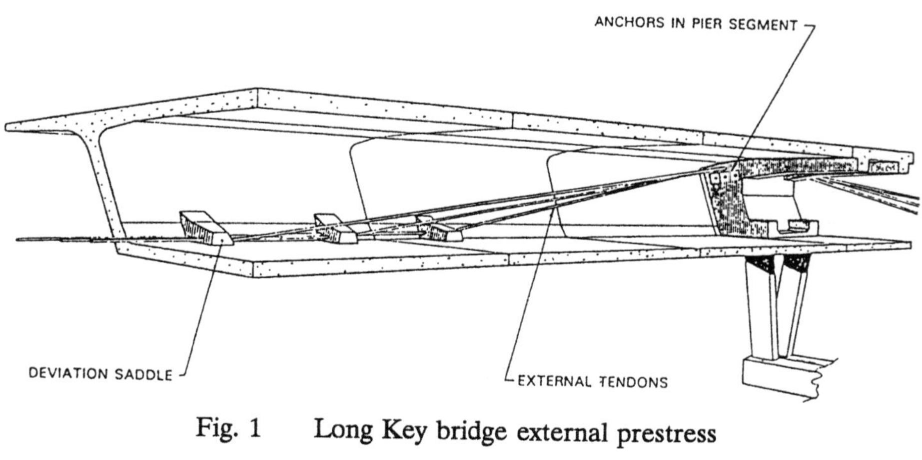


Fig. 1 Long Key bridge external prestress

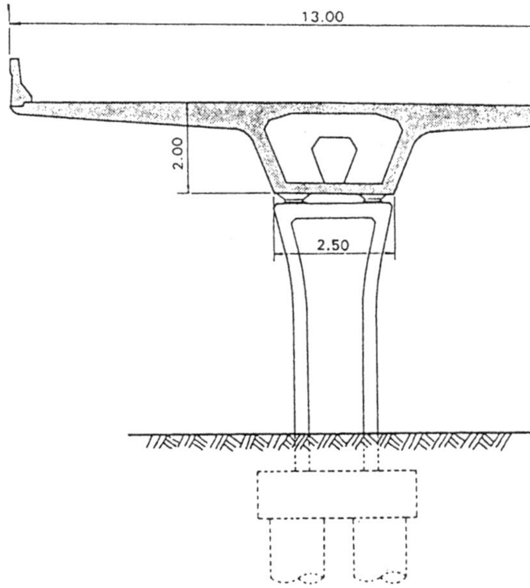


Fig. 2 San Antonio Y Project, typical cross-section

Table 1 San Antonio Y Project. Schedule of tenders

| CONTRACT | TENDER DATE | DECK AREA M2 | SF | CONTRACT AMOUNT ($M) | BRIDGE COST ($/SF) |
|---|---|---|---|---|---|
| 1A | DEC 84 | 34,000 | 366,000 | 18 | 20 |
| 2B | JUNE 85 | 32,500 | 350,000 | 19 | 25 |
| 3A-3B | MAY 86 | 52,700 | 567,000 | 36 | 35 |
| 1C | OCT 87 | 69,700 | 750,000 | 43 | 37 |
| 3C-3D | AUG 89 | 54,000 | 581,000 | 49 | 47 |
| 2C | OCT 90 | 40,800 | 439,000 | 52 | 49 |
| | TOTAL | 283,700 | 3,053,000 | 217 | 37 average |

### 2.2 Monterrey Rail Viaduct, Mexico

This is the single largest application of precast segmental construction in the field of urban light rail systems. The guideway consists of an 18.5 km long viaduct with 619 spans of an average length of 27 m, carrying twin tracks for mass transit transportation. Horizontal curves have a minimum radius of 250 m, while the maximum longitudinal slope is generally kept at 2% (special zones have 3.5%). The typical box girder deck has a constant cross section, Fig. 3.

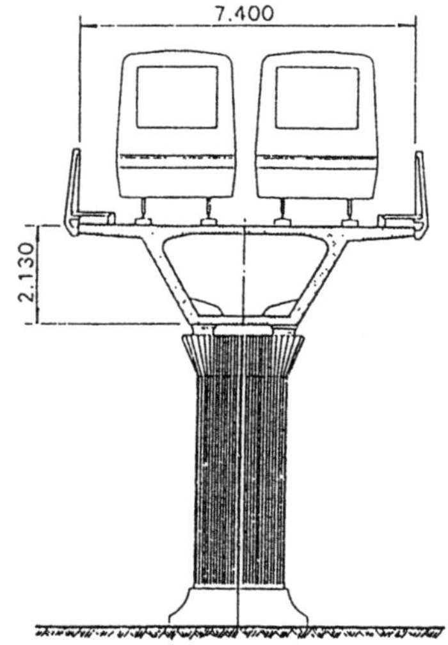


Fig. 3 Monterrey rapid transit system, typical cross-section

A total of 6500 segments were fabricated in a yard where 20 long beds were equipped with 26 moulds for typical segments and 14 moulds for pier segments. Transverse pretensioning was used in the deck slab. Segments were placed in the viaduct with 8 assembly trusses resting temporarily on pier brackets. External prestress tendons assembled the segments of each span into a monolith after matching the segments at the cold multiple-keyed joints.

The assembly schematics appear in Fig. 4. The section of a typical span was generally completed in two to three working days.

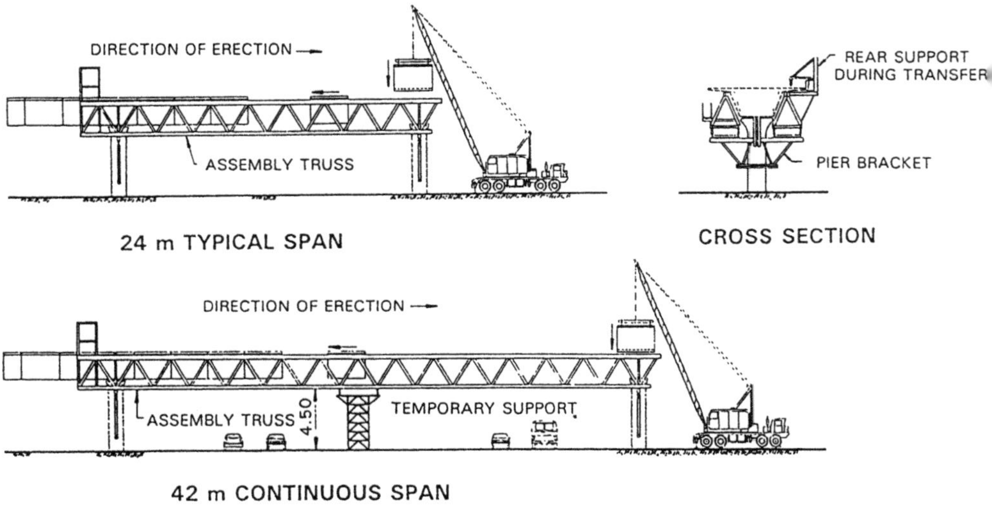


Fig. 4 Monterrey rapid transit system, assembly schematics

Noteworthy is the concept for the design and construction of the 18 elevated stations regularly located along the line at 1100 m intervals. Special post-tensioned brackets and floor beams enable the station platforms to be supported by the same typical deck cross-section, Fig. 5. Spans are reduced and the deck is made continuous over the length of the station to keep the stresses in the box girder within allowable values. The result is aesthetically pleasing whether seen from the outside or from platform level.

The entire project, including design and construction, was completed in two and a half years; construction proper was completed in less than two years. The first segment was cast in October 1988, while the last of the 619 spans was assembled in July 1990.

## 2.3 Bangkok Expressway System

When completed, this project will constitute, probably for some time, the world's largest use of precast segmental construction for an urban highway bridge project, Fig. 6. Faced with overwhelming traffic problems, the city of Bangkok has undertaken the construction of an elevated expressway system, the second

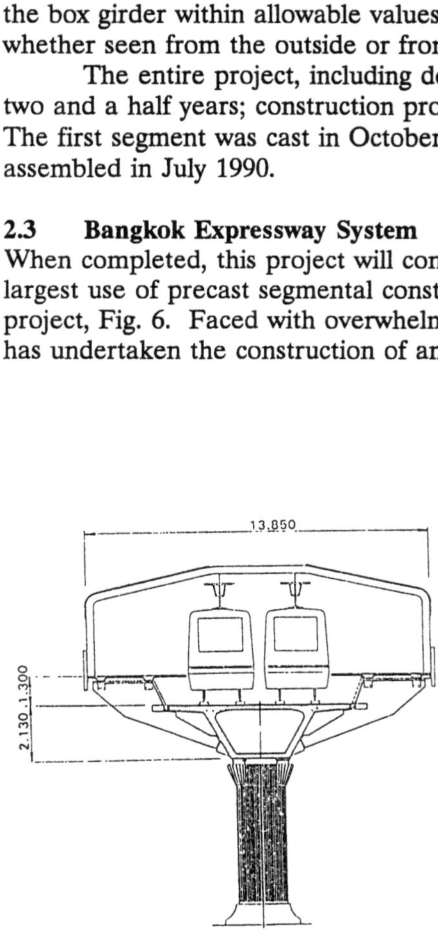


Fig. 5 Monterrey rapid transit system, typical section at stations

Fig. 6 Bangkok Expressway system, key plan

stage of which is presently in full construction (SES project). It comprises a total of 1575 spans with an average length of 42 m built using precast segmental construction techniques representing a length of 66,000 m and a deck area of 770,000 m$^2$.

The box girder deck is of a constant 2.40 m depth and has typically two different widths of 12.20 m and 15.60 m. In the transition areas, two box girders are transversely connected with a cast-in-place closure joint. The top slab is transversely post-tensioned, while spans are typically simply supported to allow for differential settlements consequent to poor and variable soil conditions.

Spans are connected longitudinally with a roadway closure in groups of four to reduce the number of expansion joints. Each span comprises 3.40 m long precast segments, assembled on cold joints with multiple keys by external tendons deviated along their trajectory and anchored in the end pier segments.

A total of 20,500 superstructure segments must be produced (14,500 for the initial part of the project and 6,000 for the incremental portion thereof) at an average of 750 segments per month thus completing the initial stage in two years.

Short-line casting was selected because of the variable geometry requirements and the limited amount of space available for the precasting yard (200,000 m$^2$). As shown on the segment precasting yard map, Fig. 7, the yard was divided into four production zones; steel cages were preassembled in a central location while concrete was produced in two batch plants.

There are 46 casting machines (32 for typical segments and 14 for pier segments). Production rates are one typical segment per day in each machine and two pier segments per week in each machine. Segment production started in February 1991 with the first two machines. In August 1991, production reached the assigned goal of 750 segments per month which was exceeded in November

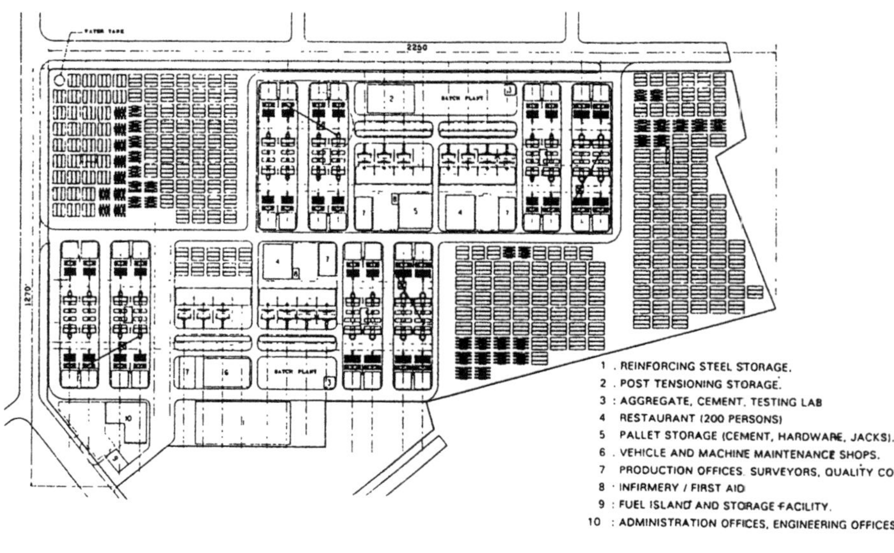


Fig. 7 Bangkok Expressway system. Layout of precasting yard

1991 with 900 segments.

A project of such magnitude presented problems of unusual interest. For example, superstructure shop drawings had to be produced in numbers and at a rate heretofore unknown. Three-dimensional CADD models were prepared and followed by almost fully-computerised drawings giving all geometry data of each span and all relevant details of each segment (dimensions, geometry control points, reinforcing steel, post-tensioning details, embedded items, etc.).

Construction of substructure works and assembly of the segmental concrete superstructure were divided into four sections with independent general contractors for each contract.

Assembly of all segmental superstructure calls on the span-by-span technique, with assembly trusses working underneath the deck, or above the deck, and segment handling from the top in both cases to reduce traffic restrictions at street level.

## 2.4 An innovative design of composite construction (steel and concrete) for medium spans: the Roize Bridge

An attempt was made to optimise the use of structural steel and high-strength concrete into a new structural scheme combining minimum weight, maximum rigidity and lower cost. A small bridge was selected for this purpose near Grenoble, France on a grade separation of the A49 tollway, Fig. 8. Basically, the deck structure comprises the following components:

(a) Precast high-strength concrete slabs (140 mm thick) carrying the highway traffic and constituting the top flange of the structural system. Design concrete strength was 80 MPa (on cylinders at 28 days). However, an actual strength of

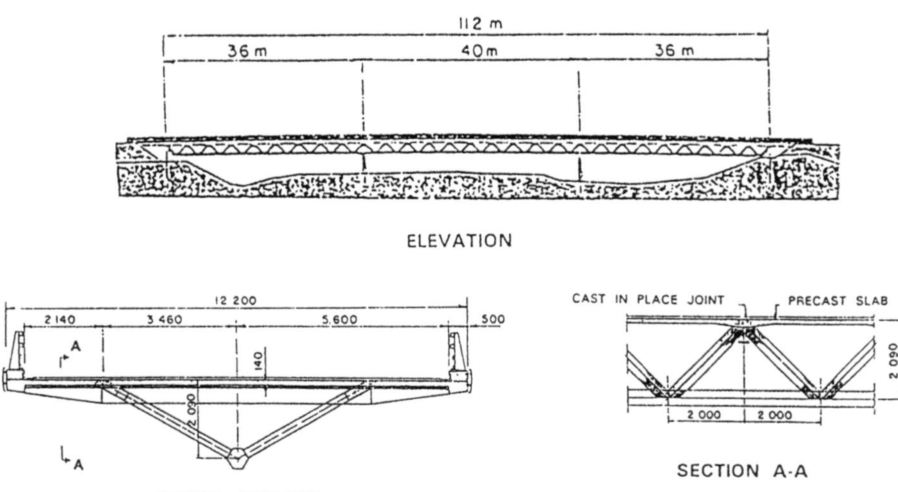


Fig. 8 The Roize bridge. Dimensions

100 MPa was reached (with 45 kg of silica fume, 450 kg of cement per $m^3$ and a water/cement ratio of 0.29 made possible with plasticizer). Concrete slabs are longitudinally pretensioned and transversely post-tensioned.
(b) A series of steel floor beams at 4 m intervals carrying the concrete slabs and made composite therewith.
(c) A space steel truss including a single bottom flange pipe connected to the floor beams by a series of diagonals located in two inclined planes with a single Warren configuration in each plane. This space system is therefore designed as the simplest assembly of individual triangles where all loads applied are basically resisted by axial loads in all structural members with minimum bending stresses.
(d) The structural capacity of the truss is complemented by external post-tensioned tendons deviated along their trajectory in much the same fashion as in a precast concrete segmental box girder design.

Design-wise, the structure is very light (approximately half the weight of an optimised concrete box of the same dimensions), while remaining rigid under live loads. Construction-wise, the deck is segmentally fabricated and assembled span by span on a truss.

A comprehensive test programme was performed on a full scale load model comprising four bays of the deck, which entirely proved the adequacy of the design on the following critical points: (i) fatigue of thin deck slab; (ii) adequacy of the connections between individual steel members and deck slab between steel and concrete; (iii) response of the structure to bending, shear and torsion.

The potential applications of this new design are for repetitive medium-span viaducts or light rigid decks for very long cable-stayed spans. Among further refinements, present studies include concrete filling of the bottom flange steel pipe and simplification of the shop and field connections between steel members.

## 3 Long-span girder bridges

Precast segmental construction has been used extensively in long span bridges over water or deep valleys for many years. The author was privileged to be associated with many such projects, particularly in the USA, including the Albemarle Sound Crossing in North Carolina and the Wando River Bridge near Charleston, South Carolina. Other outstanding applications are mentioned below.

### 3.1 H-3 Windward Viaduct, Hawaii, USA

This bridge is located on the H3 Freeway north of Honolulu on the island of Oahu. It will connect the northern shores of Kailua and Kaneohe with Pearl Harbour across the mountain ranges separating the north and south coasts of this beautiful island. Twin bridges carry the two separate traffic lanes on a total length of 2040 m, Fig. 9.

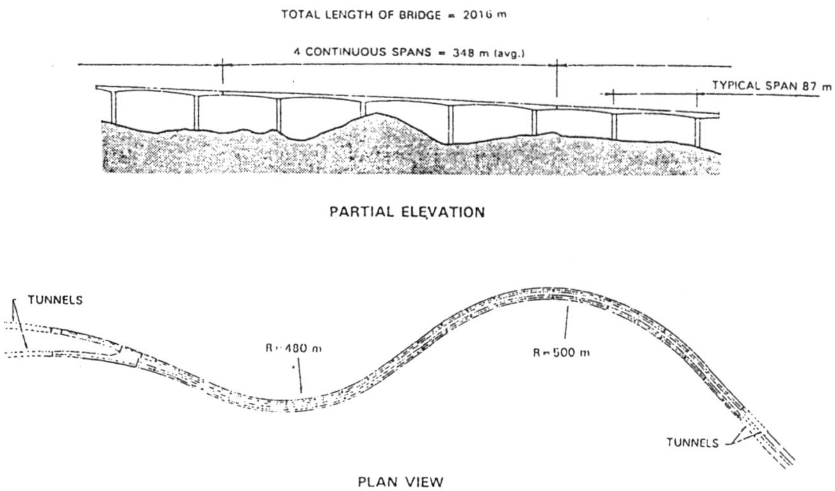


Fig. 9 H-3 Windward Viaduct, Hawaii. Partial elevation and plan view.

Initially the design featured a cast-in-place cantilever deck and cast-in-place box piers resting on large precast concrete pile foundations. The contractor's redesign (prepared by JMI) offered two major items of interest: (1) use of drilled shaft foundations enabling a substantial reduction of the impact on the protected environment (2) a precast concrete superstructure generally made integral with the piers to improve seismic behaviour. While enabling a faster construction schedule, precast segmental construction also reduces the amount of work to be performed at the construction site. Unfortunately, external prestressing was not allowed on the project and it was not possible to derive maximum benefit from the method.

The most interesting aspect of the project relates to its exceptional geometry requirements. The twin bridges follow different trajectories with a very significant longitudinal slope (5% average on the overall length), variable curves in plan (different between the two bridges) and related variations of the transverse superelevation.

Segment prefabrication was naturally handled with short line casting to accommodate the geometric constraints. The precasting yard was set up 15 miles from the construction site with four casting machines. Segments are trucked to the bridge site where they are handled with a special launching gantry, Figs 10 and 11, which enables simultaneous construction of the twin bridges in spite of the geometric variations of their respective layouts. This new patented launching gantry has proved to work very successfully. Basically it consists of the following components:

(a) A series of cross beams attached to the concrete deck which provide the successive bearings of the launching gantry proper while compensating for the transverse superelevation.

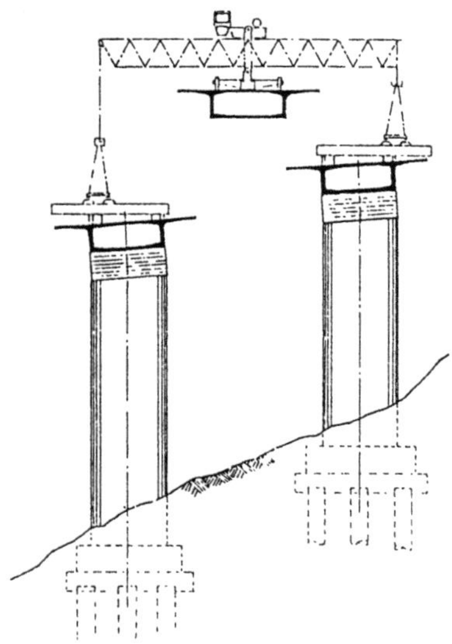

Fig. 10 H-3 Viaduct, Hawaii. Cross-section of twin viaducts

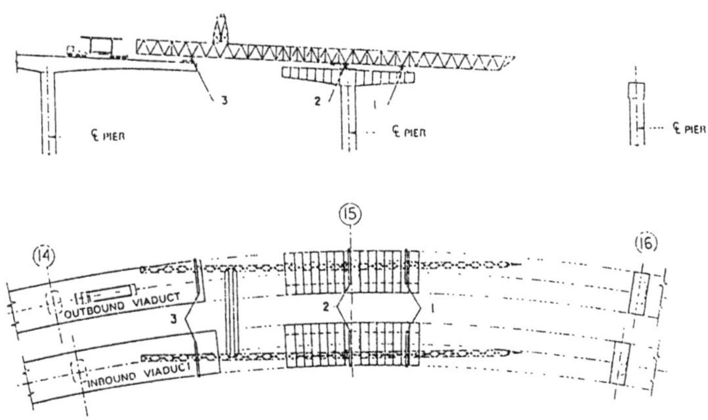


Fig. 11 H-3 Viaduct, Hawaii. Launching gantry schematics

(b) Two longitudinal parallel steel trusses resting on at least two of the cross beams. One truss is located above the downstream web of the downstream bridge while the other is above the upstream web of the upstream bridge. Such configuration clears the way for the segments to be placed onto either bridge without moving the launching gantry crosswise.
(c) A portal crane travels the steel truss, picking up the segments at the rear from the finished bridge to place them into their final position in balanced cantilever; this portal crane has one fixed leg (except for a rotational capacity around a vertical axis) above the upstream bridge while the other leg with

adjustable vertical height has the necessary degrees of freedom to adjust to the bridge geometry.

(d) A very simple temporary attachment of the portal crane to the concrete bridge enables the two longitudinal trusses to be launched separately to the next pier. In fact, because construction proceeds downgrade in the direction of the longitudinal slope, advancement of the launching gantry consists rather in safely holding back the truss from sliding forward spontaneously.

### 3.2 Location of expansion joints in long-span cantilever bridges

This is an important design aspect which has long-term consequences on the performance and rideability of long-span concrete bridges.

Historically, the first prestressed concrete cantilever bridges were provided with an expansion joint at the centre of each span with a shear transfer connection allowing free longitudinal movements while maintaining the same vertical displacements of the two adjacent cantilever arm tips. Experience showed these structures to be very sensitive to long-term material properties (steel relaxation and concrete creep); large permanent vertical deflections occurred at the centre of each span accompanied by significant angle breaks objectionable from the twofold aspect of aesthetics and continuity of the roadway or rail profile.

The author was privileged to design the first precast segmental cantilever bridge with continuous deck (Choisy-le-Roi). The individual cantilevered arms are connected at each span centreline with continuity tendons ensuring the performance of a monolithic structure. The problem of long and short-term deflections and angle breaks is provided with a final solution.

However, other aspects of the effect of concrete creep and thermal gradients later appeared in other such continuous cantilever bridges. Significant tensile stresses develop on the bottom flange near the span centre due to the combined effect of moment redistribution and temperature gradients. Design assumptions had to be revised to properly account for these factors which were originally ignored or under-estimated. Today, most long-span cantilever bridges have continuous decks, and realistic design assumptions now ensure their excellent performance.

There remains, however, the problem of expansion of long decks (say above 300–400 m long). Where should the expansion joints be located along the span when needed? The answer appears clearly while considering the graph in Fig. 12 which shows the various deflection curves under live load for different locations of the hinge along one span of the Oleron Viaduct in France. When the hinge is located at the point of contraflexure of dead load moments (close to the quarter span point), there is no difference between the continuous and hinged structure under dead load, while the live load angle break and vertical deflections are significantly reduced. Consequent to these studies, it was decided to generalise the use of a quarter span in lieu of a centre span hinge. Long bridges such as St Cloud in Paris, Sallingsund in Denmark, Columbia River and Zilwaukee Bridges in the USA were designed accordingly. More recently, the H-3 Viaduct in Hawaii follows the same principle. However, it must be recognised

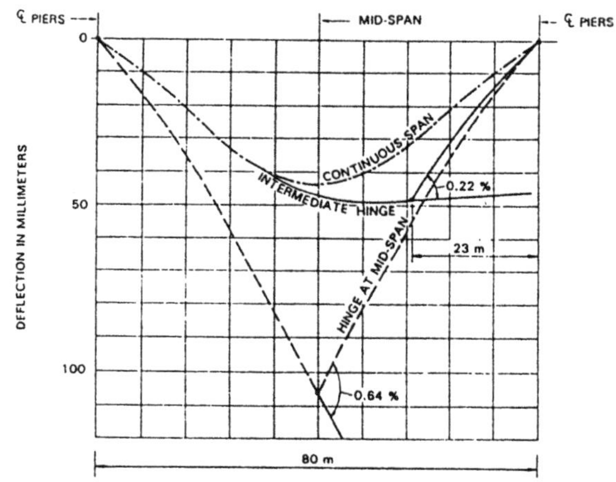


Fig. 12 Oléron Viaduct. Deflection curves under live load

that the desired performance for rigidity and geometry is attained at the expense of construction complexity. In particular, it has never been easy to carry the cantilever construction process through the hinge segment which needs to be temporarily blocked; moreover, the layout of longitudinal prestress in the hinged span is never a model of simplicity. The newer concept proposed below should fully answer this need for improvement.

According to the scheme outlined in Fig. 13, the expansion joint is moved back to the span centre, where an active connection is provided between the two adjacent cantilever arms for the purpose of transferring at all times (long-term under dead load, and short-term under live load), a shear force and a moment in an active fashion; in other words, the magnitude of shear load and moment across the joint may be monitored at the designer's request to restore all the geometric and mechanical properties of a continuous deck across the expansion joint.

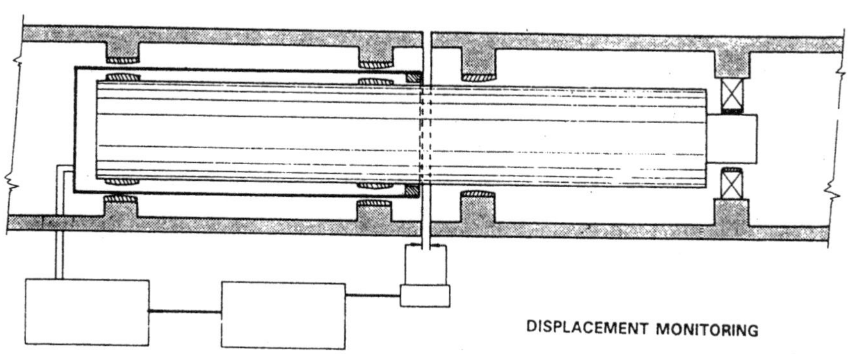


Fig. 13 Expansion joint with active connection

Studies made on a bridge in the design stage in France are summarised in Figs. 14 and 15. The conventional centre hinge with shear transfer only exemplifies the usual large deflections and angle breaks while the active connection with shear and moment transfer shows the problem of deck geometry to be under full control. A continuous member across the joint rests on two diaphragms both sides therefrom and may be adjusted before the bridge is opened to traffic to totally compensate all subsequent effects of steel relaxation and concrete creep.

In fact, this is only a first step and one may easily imagine the extension of the foregoing principles into a fully monitored system which could control the structure geometry under all exterior loads (traffic, wind, earthquakes, etc) at all times. This is an attempt at using artificial intelligence in civil engineering structures.

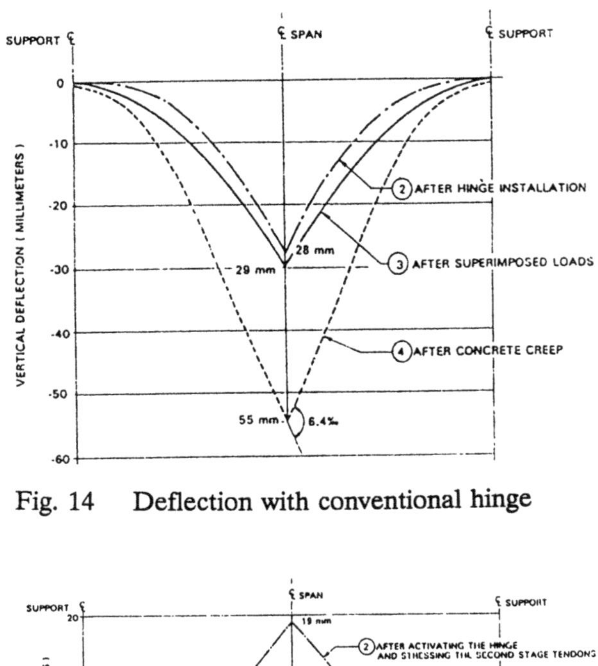


Fig. 14 Deflection with conventional hinge

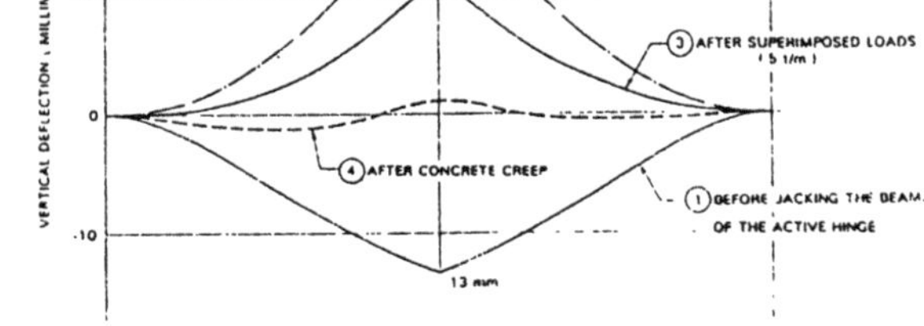


Fig. 15 Deflection with active hinge

### 3.3 Long-span trusses

Another area of interest in long-span concrete (or composite steel and concrete) bridges lies in the potential of trussed structures.

Cantilever trusses were abundantly used at the turn of the century for many outstanding steel structures; in recent decades, however, other types of designs such as orthotropic steel box girders became more popular, particularly outside the USA.

It is firmly believed that the association of high strength concrete and structural steel trusses could produce excellent and economical designs in the field of long span bridges.

To demonstrate the potential of this approach, a preliminary design was prepared for the crossing of a deep valley in South Eastern France, with free spans of 467 m. Figs. 16 and 17 show the elevation and typical cross-section of the project.

The 21-m wide deck is made of a high-strength prestressed concrete slab resting on three vertical steel trusses. The bottom chord is also of high strength concrete on either side of the piers over one third of the span length, while the middle third is of steel only to reduce the actual weight of the structure. Deck construction will be by balanced cantilever in increments of 15 m (distance between nodes on top and bottom flanges).

Diagonals are first placed and connected to a movable form at top and bottom flange levels to allow for concrete casting. The amount of longitudinal prestress over the piers to resist the cantilever moments is such that only some tendons can be placed in the concrete slab while the balance would be external below the deck slab.

Cost estimates show this structure type to be competitive with a cable-stayed design having a centre span of 860 m (see section 4.4).

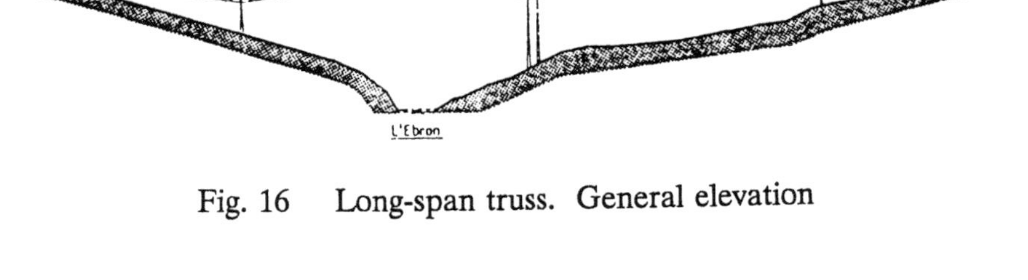


Fig. 16 Long-span truss. General elevation

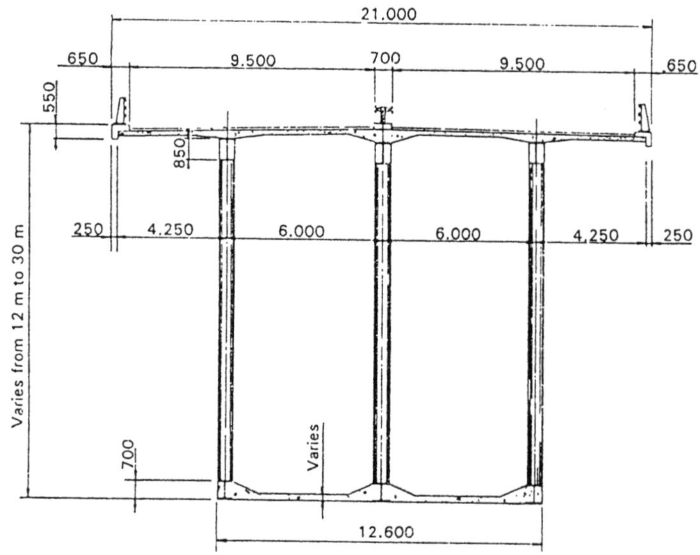


Fig. 17 Long-span truss. Typical cross-section

## 4 Cable-stayed bridges

In recent years, many interesting cable-stayed structures have been designed and built around the world. A few of these with which the author has been closely associated will be mentioned in this overview.

### 4.1 Twin box girder decks with centre suspension

The first application of the scheme was for the design and construction of the James River Bridge south of Richmond, Virginia, USA, which opened to traffic in 1990. The 45-m approach spans are continued up to the main piers where the pylon tower is located between the typical piers of the twin box girders.

In the stayed spans, the transfer of loads between the box girders and the centre suspension is achieved by a series of inside struts (in each box girder) and precast concrete delta frame units between the girders working together with the transversely prestressed centre closure deck slab.

An improved scheme was later developed as shown in Fig. 18. The stay load is transferred in equal parts to the top and bottom slabs to obtain a more direct distribution of the high concentrated load to the entire deck cross-section. The precast concrete delta frame which is very heavily reinforced would be replaced by a prefabricated steel unit.

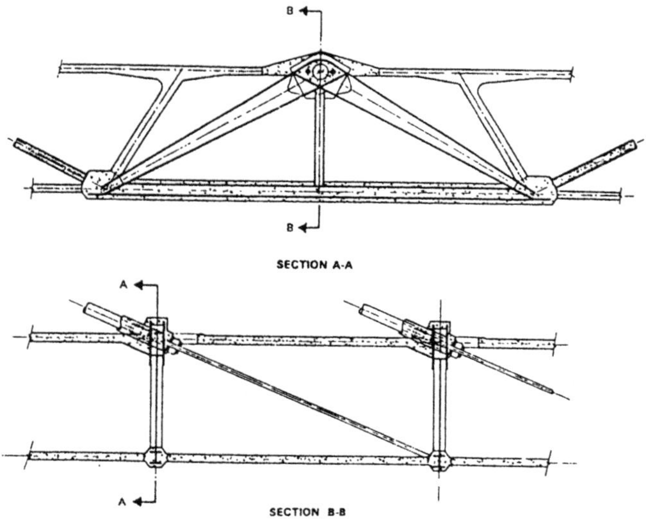


Fig. 18 James River bridge. Steel delta frame - partial longitudinal section

### 4.2 Streamlined box girder deck with twin suspension

Neches River Bridge, near Port Arthur, Texas, USA was opened to traffic in late 1990, Fig. 19, and affords an interesting example of a fully precast structure incorporating the following noteworthy features:

(a) streamlined precast box girder deck. Drag coefficient for lateral wind is only 0.70 as opposed to 1.34 for a typical box girder deck with overhangs and tapered webs (like Brotonne and Sunshine Skyway Bridges);
(b) two outside planes of vertical stays anchored in the deck's outer face, resting on precast segmental pylons;
(c) active deviation saddles located in the inner void of the boxed pylon section to allow the continuous stays across the pylon to be easily monitored and readjusted as required to the effect of concrete creep.

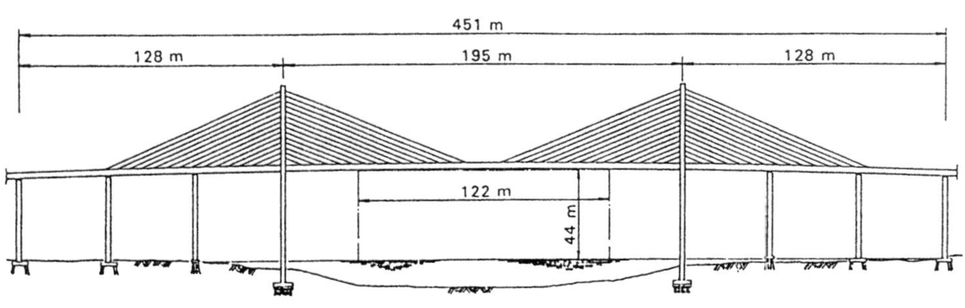


Fig. 19 Neches River Bridge. General elevation

**4.3 Isère River Bridge on A49 Tollway between Grenoble and Valence, France**
This important cable-stayed structure was opened to traffic in December 1991 to carry the new 4-lane tollway over the deep valley of the Isère River. A single pylon was located along the right bank of the river while avoiding the rugged and somewhat unstable left bank. Two short piers are added in front of the end abutments, Fig. 20, to best control the roadway profile under traffic.

Another intermediate pier is added in the right bank mainspan to reduce deck bending under live load (the French code is rather aggressive in this respect) while providing additional support to the pylon deck system during construction under variable wind loading.

The deck cross-section, Fig. 21, provides the ultimate streamlined shape

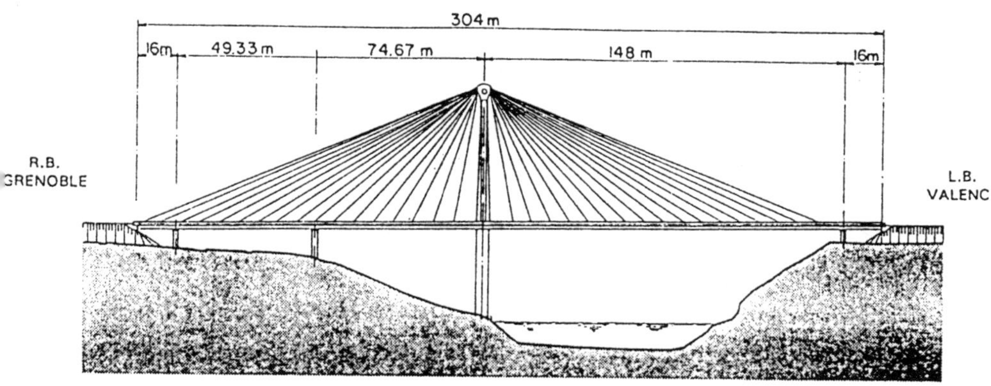


Fig. 20 Isère River bridge. General elevation

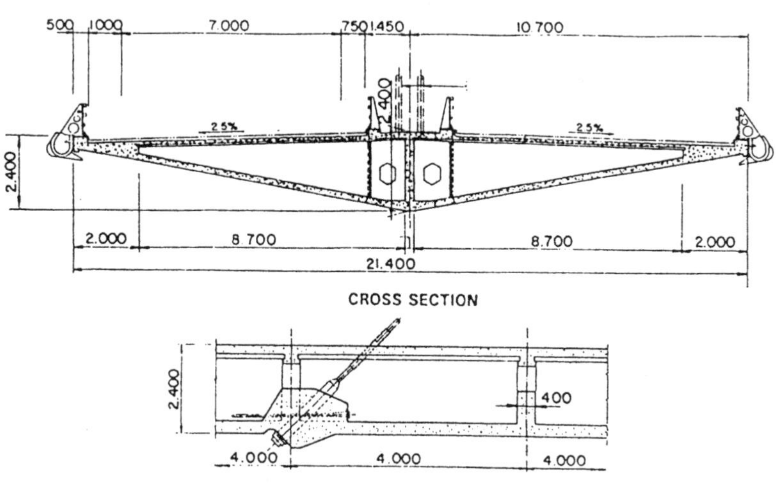


Fig. 21 Isère River bridge. Deck cross-section and partial longitudinal section

formed by two triangles abutting on a central longitudinal web. Stays are placed at 8-m intervals and are located in pairs on either side of the centre web with anchor heads located at deck soffit level.

The reinforced concrete pylon, Fig. 22, comprises a boxed shaft below the deck extending thereabove to a concrete saddle on which all continuous stays are deviated. The cruciform section varies continuously between saddle and deck levels to follow the stress flow due to weight, longitudinal live load bending and transverse wind bending.

The deck is a cast-in-place cantilever construction with the centre web cast first to allow installation of the stays, followed by the two triangles which are cast on suspended forms. Architectural details have been incorporated into the design shapes of the pylon to improve its appearance at limited additional expense.

## 4.4 Very long-span cable-stayed bridges with steel space-trussed deck

The space truss concept developed for the construction of the Roize Bridge (section 2.4) could be remarkably extrapolated to produce an efficient deck for long cable-stayed structures.

A tentative application was studied for the Ebron Crossing with a main span of 860 m, Fig. 23. The deck typical cross-section is shown in Fig. 24. There are two planes of stays on the edge of the deck with stays located at 20 m intervals. Surprisingly, the results of a scale wind model test show that the truss

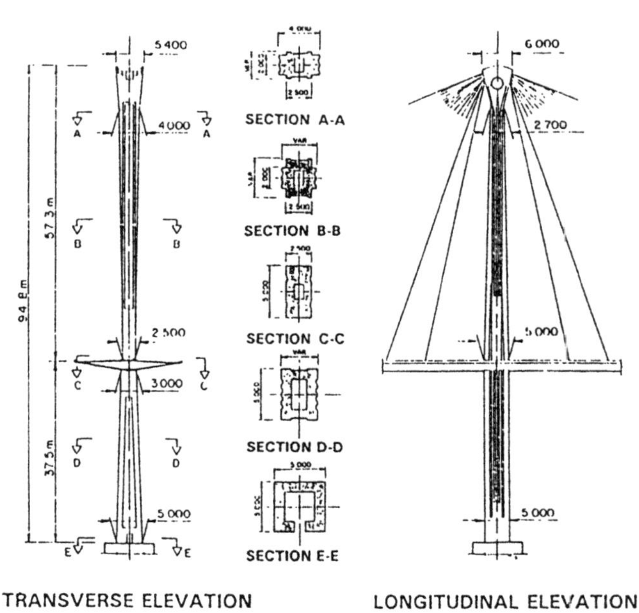


Fig. 22 Isère River bridge. Pylon dimensions

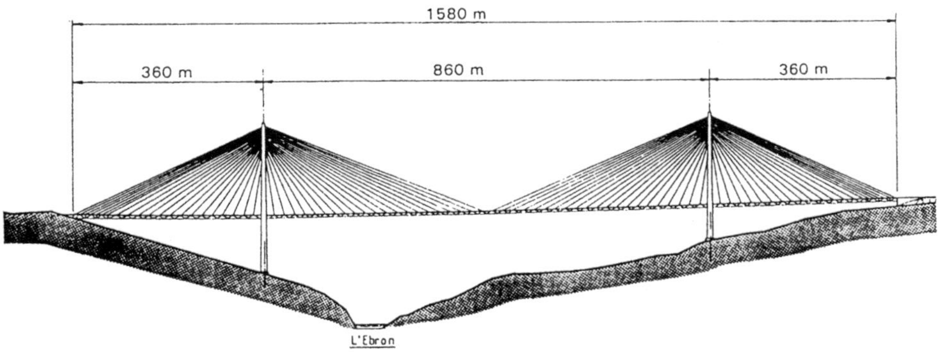


Fig. 23 Ebron Crossing. Elevation

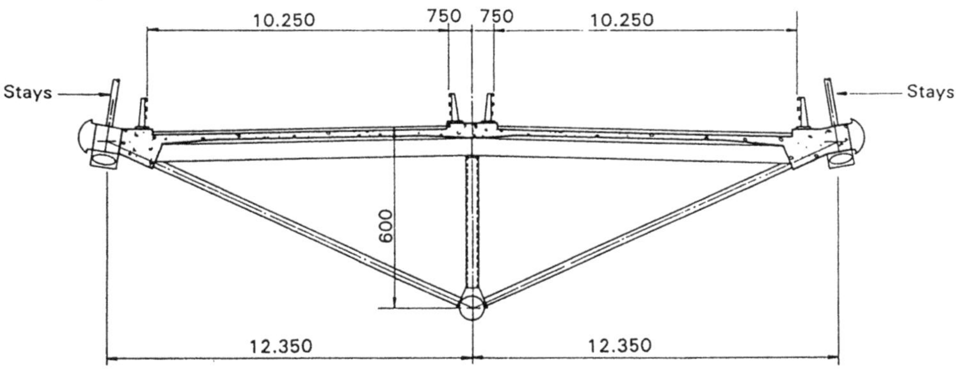


Fig. 24 Ebron Crossing. Typical cross-section

has a drag factor less favourable than a streamlined box with continuous surfaces To minimise the wind effects on the deck, it will be necessary to install buffers outside the deck edges to guide the airflow away from the members of the space truss. Cost estimates show the 860 m cable-stayed scheme to be equivalent to the composite, 467 m truss scheme (see section 3.3).

### 4.5 The bi-stayed concept

All bridges with a clear span longer than 500 m have been built to date as suspension bridges. Currently, the world's longest span is in the design and construction stage in Japan, with a clear span of almost 2000 m. Meanwhile, the longest cable-stayed bridges are Annacis Bridge in Canada with a clear span of 465 m and the Normandy Bridge in France, with a clear span of 856 m.

With existing materials (steel with a yield strength of 350 MPa and high strength concrete of say 100 MPa), the maximum span of a conventional cable-stayed deck lies between 1000 and 1500 m. The critical factor is the magnitude of compressive stresses in the deck around the pylon consequent to the cumulation of the horizontal component of the stay loads. Using a radiating stay configuration reduces this load to half the value reached with a harp-shaped layout (all stays parallel).

To increase the clear span beyond the limit which corresponds to the saturation of the deck section with axial compression stresses, it is necessary to find a static scheme enabling any additional loads to be supported without increasing the deck stresses. The principle, Fig. 25(a), is to suspend the main span deck loads by two different families of stays (whence the expression bi-stayed bridge); travelling from the pylon towards the span centre line, one will find:

on a length a1 (conveniently the length of the side span), conventional stays (S1) symmetrical with regard to the pylon centreline are self-anchored into the deck and create a maximum axial load N1;

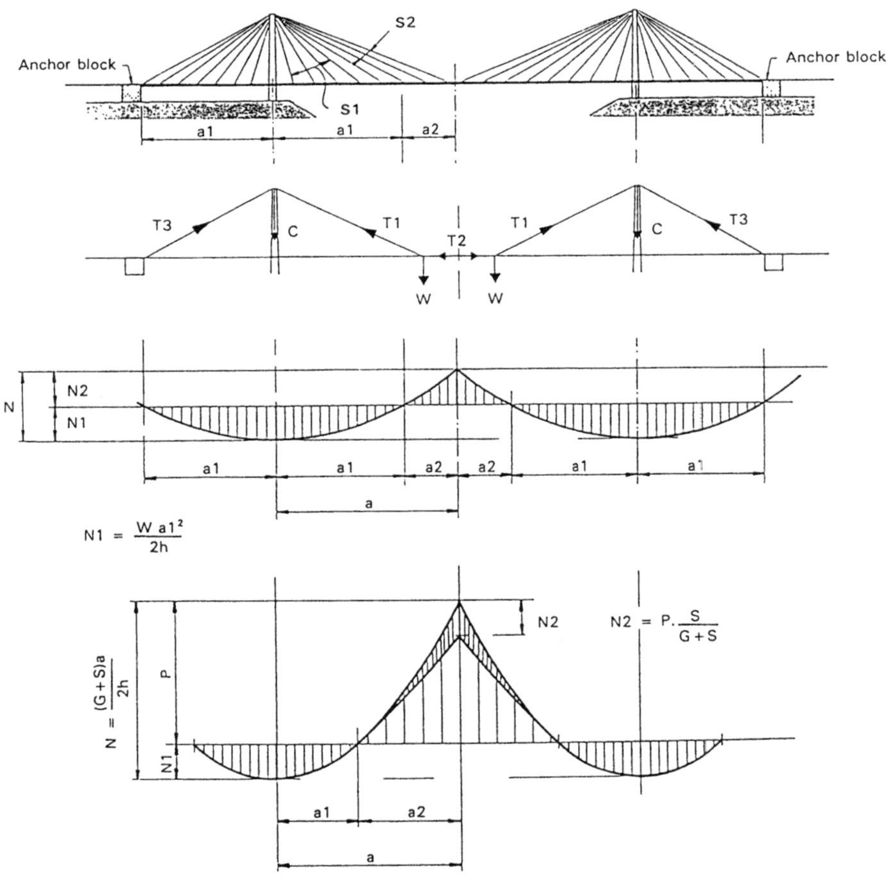


Fig. 25 Bi-stayed bridge principles

on a length a2, another family of stays (S2) which are now symmetrical with regard to the span centreline (and not the pylon centreline as they appear in Fig. 25(b)), to balance the successive pairs of unit loads W in developing a tensile axial load T2 in the centre area of the main span rather than a compressive axial load near the pylons. To ensure equilibrium, the stays must be deviated above the pylons and anchored away from the deck in an earth abutment similar to the anchor blocks of the conventional suspension bridges.

The diagram of deck axial load along the span which appears in Fig. 25(c), shows that the tensile axial load N2 will add to the compressive axial load N1 to balance the total horizontal component of stay loads for the full half span $a = a1 + a2$. For the same material properties, it is therefore possible to theoretically increase the maximum clear span of a cable-stayed bridge in the proportion of $\sqrt{2} = 1.41$ and therefore a 1500 to 1800 m span should become feasible. However, one drawback of the system lies in the fact that heavy permanent tensile stresses exist in the deck at mid-span.

There is a way to compensate for these tensile stresses while simultaneously increasing the maximum span very significantly. Longitudinal post-tensioning tendons will be installed in the deck to create a load P which compensates for the total axial load created by dead and live loads, Fig. 25(d).

When the deck is fully loaded, there is a neutral state of axial stress, and when the deck is unloaded (dead loads only) the axial load is reduced to a compression load $N2 = P.S/G+S$ (G is dead load and S Live load). Therefore $N = N1 + P = N1\ (2 + G/S)$.

For a long span highway bridge, G/S is equal to about 3 and $N = \sqrt{5}N1$. Consequently, the limiting span is increased in the proportion 2.2. This principle makes a clear span of 2500 to 3000 m now feasible.

A remarkable advantage of the proposed scheme is that a full balance of all dead loads can be attained in terms of deck axial load. In particular, the deck remains insensitive to the rapid increase of the stay weight in very long spans.

Many practical aspects must be considered in the implementation of the new concept, such as aeroelastic stability and constructability. To preserve the advantages of the method, cantilever construction should be used for the entire main span.

A design was prepared according to the bi-stayed concept for a 1200 m clear span carrying a combination of six highway traffic lanes and two trains. Fig. 26 shows the elevation and Fig. 27 the typical cross-section. Near the pylon, the box section is entirely of concrete while a composite concrete/steel section is used on a length of 800 m in the centre span.

Fig. 28 shows the comparison of the deck axial load during construction and in the finished bridge. There is only a 15% increase (22,300 t versus 19,300 t) of deck axial load during construction, which is acceptable.

It is interesting to compare the vertical deflection responses of a bi-stayed bridge under live load with those of a conventional suspension bridge. For the example mentioned above, subjected to a highway loading of 3.4 t/m and a train loading of 8.4 t/m, the two deflection curves are shown in Fig. 29. The superiority of the bi-stayed bridge is evident.

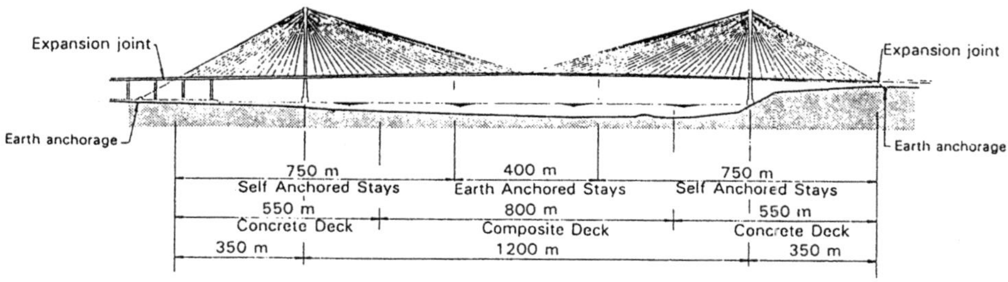


Fig. 26 Bi-stayed bridge. Elevation

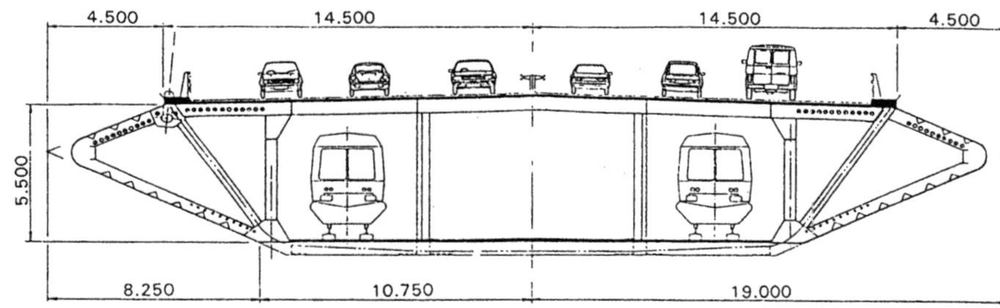


Fig. 27 Bi-stayed bridge cross-section

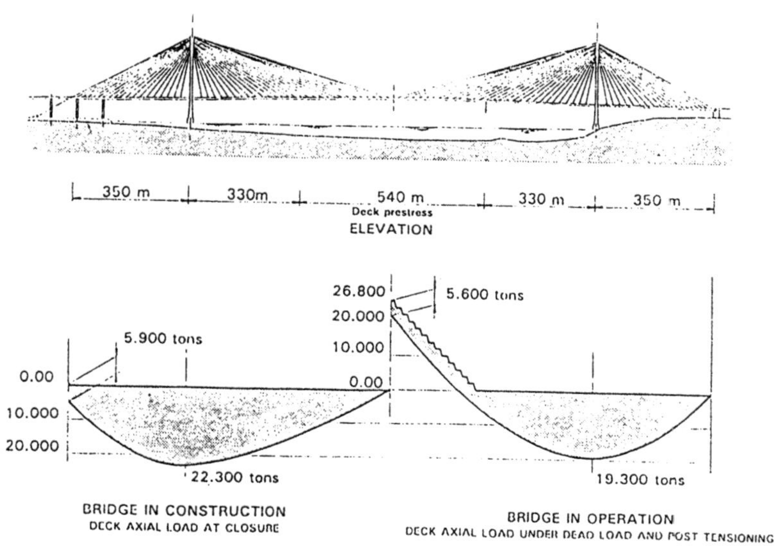


Fig. 28 Bi-stayed bridge. Comparison of deck axial load

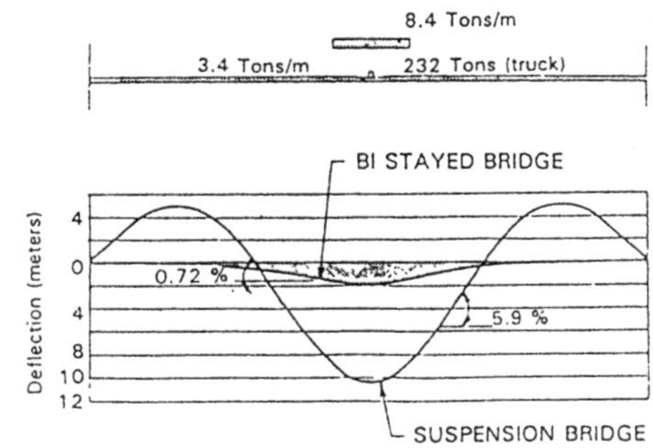


Fig. 29 Bi-stayed and suspension bridges. Comparison of deflections

To derive the full benefit of the bi-stayed concept, there are new features which need to be seriously considered:

(a) vary the deck design along the span by associating the materials in their most efficient use;
(b) in particular, use of a composite slab design comprising a stiffened steel plate (used during construction), subsequently covered with high-strength concrete for the structure in operation;
(c) use higher allowable stresses in the stays beyond those accepted by most current codes which limit the maximum stress to 45% of the ultimate strength without scientific basis;
(d) use compact stays to reduce to the absolute minimum the cross-sectional area exposed to wind, because in very long-span cable-stayed bridges, wind on the stays is the major component of the total wind load.

## 5 Conclusions and future prospects

Precast segmental concrete has had an interesting history, and still has a bright future. The method proves particularly suitable for building highway and mass transit structures in difficult urban environments.

Cable-stayed structures have increased in popularity worldwide in recent years. In the range of spans where box girders can be used (300 m for solid webs and at least 500 m for open webs with steel diagonals), cable-stayed bridges will not replace the other designs until drastic changes take place both in design assumptions and construction requirements to reduce the weight and cost of stays, which represent today a major component of the construction expenditure.

At the other end of the spectrum, the new bi-stayed bridges in a field heretofore reserved for suspension bridges, enable more rigid and more stable structures to be built at a lower cost.

Recent developments in composite structures using the space truss concept look promising and should open a new field of applications either in medium span long viaducts or in long span cable stayed bridges.

# 12 ITALIAN JOINT-FREE BRIDGE DECKS

F. MARTINEZ Y CABRERA
Department of Structural Engineering,
Polytechnic of Milan, Italy

Abstract
This paper discusses the state of art, in Italy, of joint-free bridge decks.
After some considerations about the design and the construction systems, the paper analyzes some of the most important italian bridges.

## 1 Introduction

In the last ten years in Italy a shift has occurred toward the adoption of a continuous beam scheme for motorway viaducts, using both constant and variable cross sections.

This shift has been favored by the remarkable evolution of the bearings. Moreover, bridges without joints present several advantages in the seismic zones.

Some disadvantages of the continuous beam can be summarized as follows;

- stresses induced by prestressing, temperature variations and settlement of foundations. With an appropriate monitoring process, however, it is possible to know the load variations on the bearings due to the above mentioned actions. The present technology allows one to counteract possible settlements of the piers with variable settings of the bearings (autoleveling bearings);
- presence of prestressing cables in the deck,which weaken the deck in sections with negative moments. The cables are exposed to corrosion processes if the deck becomes cracked or if the impermeabilization process is not correctly executed;
- high prestress losses due to friction for long cables. If the length of the cable is reduced, the number of anchorages is increased and this require the solution of many local static problems.

The type of continuous viaducts built in Italy in the last years may be divided into three classes:

Class A: Viaducts with solid or hollow cross section (in general a single cell) built in place, span by

*Continuous and Integral Bridges.* Edited by B.P. Pritchard. Published in 1994
by Taylor & Francis, 2 Park Square, Milton Park, Abingdon, Oxon, OX14 4RN. ISBN: 0 419 19030 9.

span. Mobile launching steel girders are employed. The girder can have a costant or variable cross section (Figs, 1.1, 1.2).

Class B: Viaduct with precast beams mounted on adjacent piers and connected through post-tensioned systems. The cross section may be constant or variable. These prefabricated elements of large dimensions are mounted with launching steel girders and progressively connected to the preceding span.

The author has designed some viaducts dividing longitudinally the box beams in two halves. Each half has been precast separately, and after the positioning it has been connected in the trasversal and longitudinal direction.

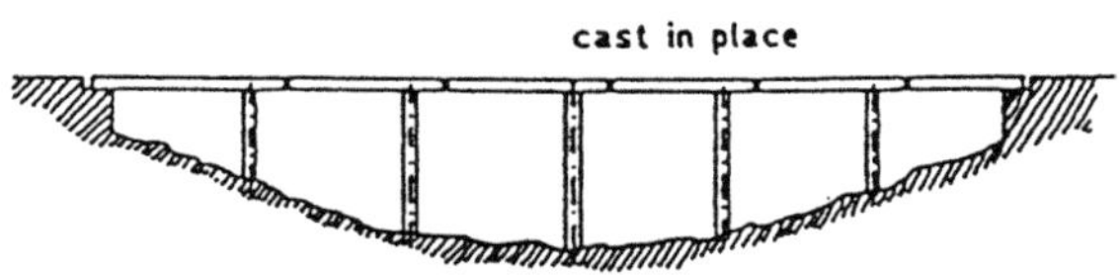


Fig. 1.1

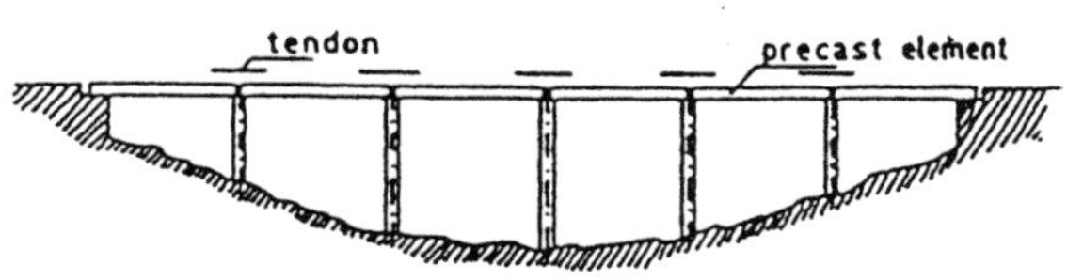


Fig. 1.2

Class C: Cantilever construction from pier with box-girder section (in general a single cell).
The segments may be precast or cast in place, and with constant or variable section (Fig. 1.3)

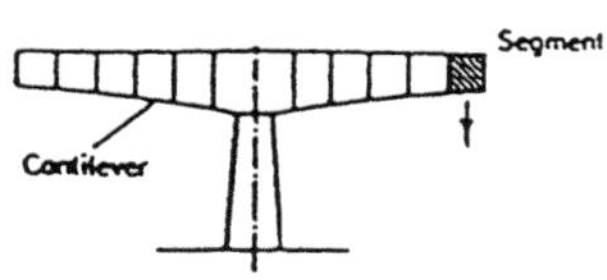


Fig. 1.3

## 2 The design

It is well known that the design and the calculations of continuous beam viaducts in prestressed reinforced concrete have many complications due to the hyperstatic bending moments induced by prestensioning and due to the progressive variation of the static scheme.

It is moreover necessary to study in depth the choice of the spans. For very long viaducts it can often be convenient to choose an average repeated span and to adopt some bigger spans in special situations. In this case it is possible to adopt special segments of beam with variable height.

For example for the Restello Viaduct on the motorway Vittorio Veneto - Pian di Vedoia (that will be shown later) spans of 60 m length average and special spans of 100 m length have bean designed. These special spans are built with segments of variable height.

The design problems can be summarized in this way

- relationship between outside and inner span: $\frac{l_e}{l_i} = 0{,}75$ (Fig. 2.1)
  It is necessary to choose the outside span always bigger than 0,55 $l_i$. Otherwise the reaction of the bearings can change the sign for the loads. For $l_e < 0{,}55\ l_i$ it is necessary to some counterweight (Fig. 2.2);

Fig. 2.1

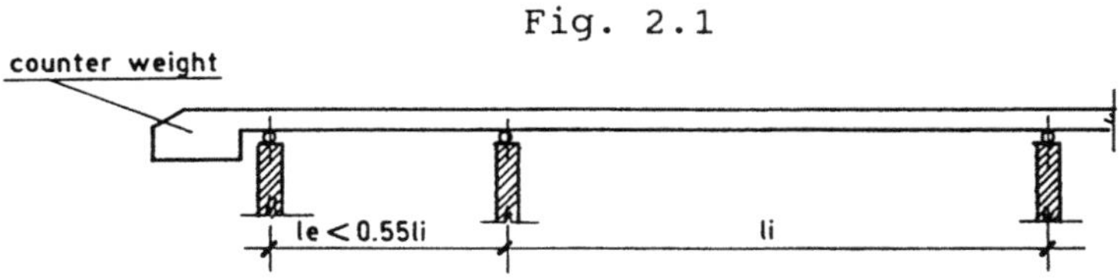


Fig. 2.2

- normally the height of the cross-section can be constant for values of the span till 60-80 m.
  Over 60 m it is necessary to increase the depth of the lower deck near the internal bearings (Fig. 2.3).

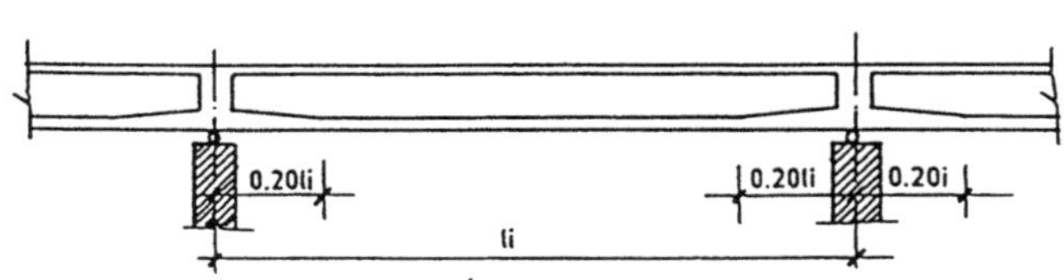


Fig. 2.3

- generally over 60 span m span it is convenient to adopt a variable longitudinal profile with a relationship between the height on the pier and the height in the middle span about 1,8 - 3,0.
  The height variation should ensure that the resultant of

the compression stresses and the tension stresses have a linear variation along the span so that it is possible to have constant depth for the webs.
For bridges with variable profile we have the following relationships: (Fig.2.4).

$$\frac{H_i}{l_i} = \frac{1}{16} - \frac{1}{20} \qquad \frac{Hm}{l_i} = \frac{1}{10} - \frac{1}{50}$$

Normally we adopte the following variation of the profile.

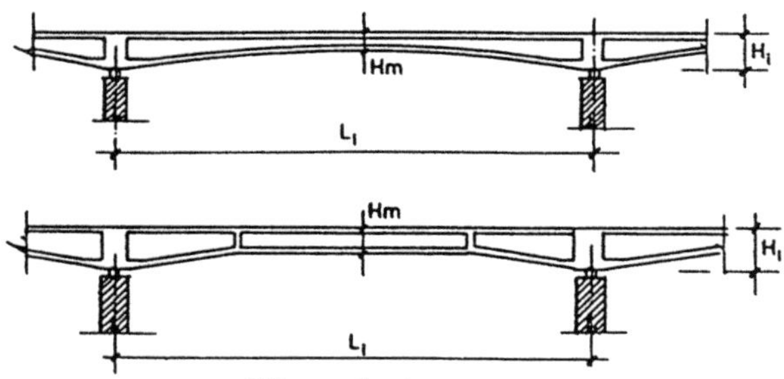


Fig. 2.4

When we have very big spans it is convenient to have double bearings so that it is possible to have an almost fixed-end beam (Fig. 2.5) (Ponte di Tadasuni).

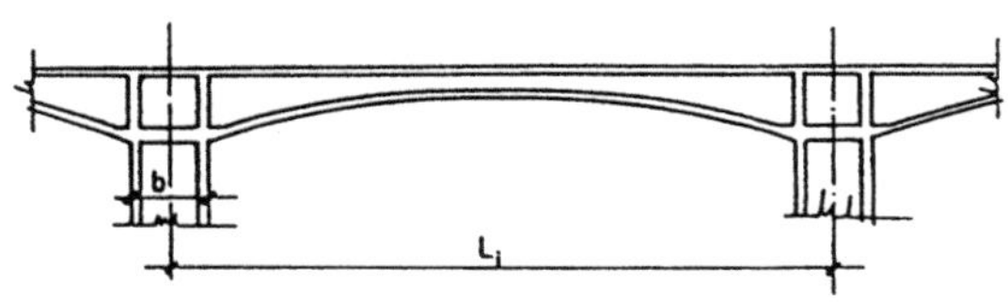


Fig.2.5

**2.1 Cross-sections adopted in Italy**
Generally the cross-sections adopted in Italy for continuous bridges are the following:

**a)** MONO-CELLULAR BOX-GIRDER (Fig. 2.6)

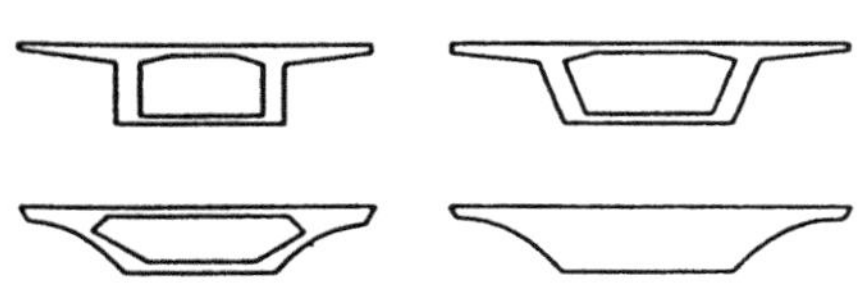

Fig. 2.6

**b)** BOX-GIRDER WITH PROTECTED CANTILEVER SLAB (Fig. 2.7)

Fig. 2.7

**c)** TWO CONNECTED BOX SECTION (Fig. 2.8)

Fig. 2.8

**d)** TWO BOX-GIRDER SECTIONS (Fig. 2.9)

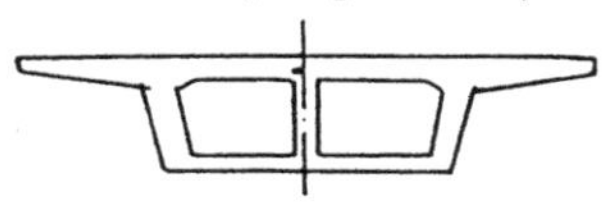

Fig. 2.9

**e)** CROSS SECTIONS CONNECTED WITH PARTICULAR PROCEDURE (Fig.2.10)

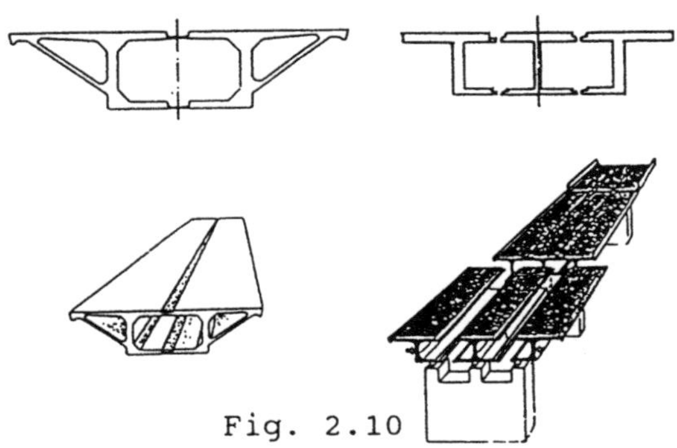

Fig. 2.10

The single segments of these cross-sections are the following (Fig. 2.11)

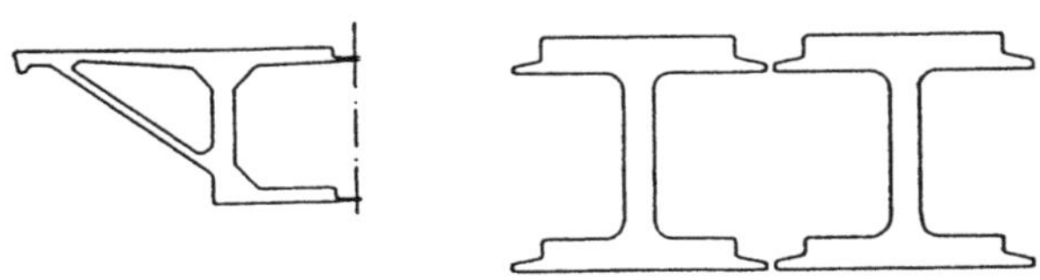

Fig. 2.11

**f)** CROSS-SECTIONS FOR PRECAST BEAMS (Fig.2.12)
(L = 20 - 32 m)

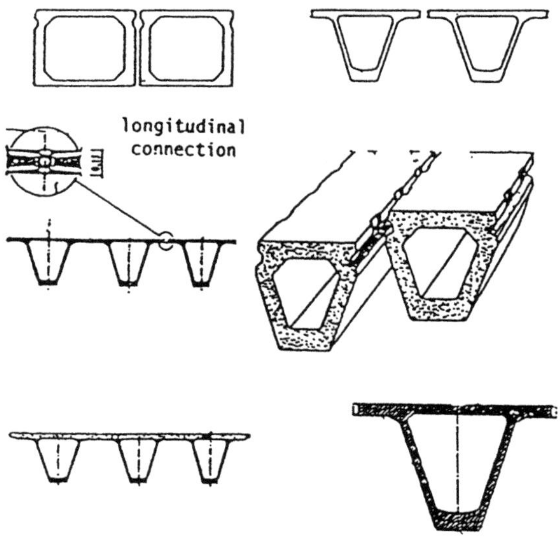


Fig. 2.12

**g)** SHELL SECTIONS (L = 20 - 25 m) (Fig. 2.13)
With this construction system we have a precast concrete shell section and we put inside it the ordinary and pre-stressed reinforcement.
Afterwards it is possible to complete the final section with concrete cast in place.
The advantage of this system is the minimum weight of the precast element.

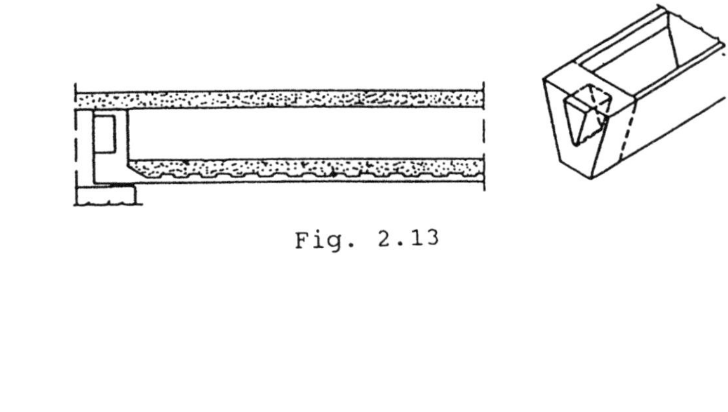


Fig. 2.13

## 2.2 Typical section (Fig. 2.14)

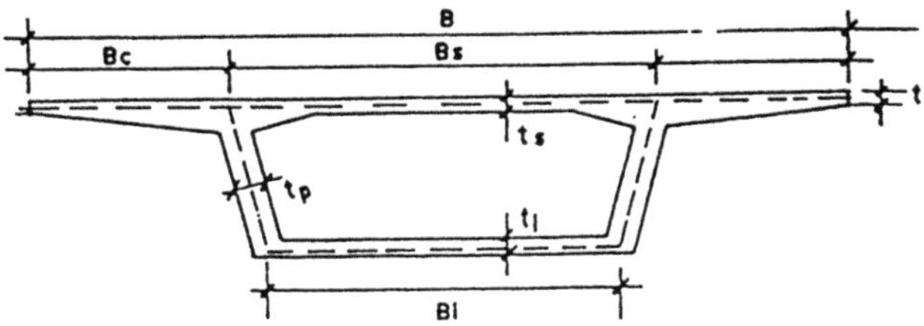


$$B_c = 0{,}40\ B_s$$

$$0{,}25 \leq t_s \leq \frac{B_s}{30}$$

$$t_i \geq 0{,}20$$

Fig. 2.14

A statistical study due to Prof. Menditto and Prof. Dezi (Ancona) shows the results reported in Fig. 2.15.

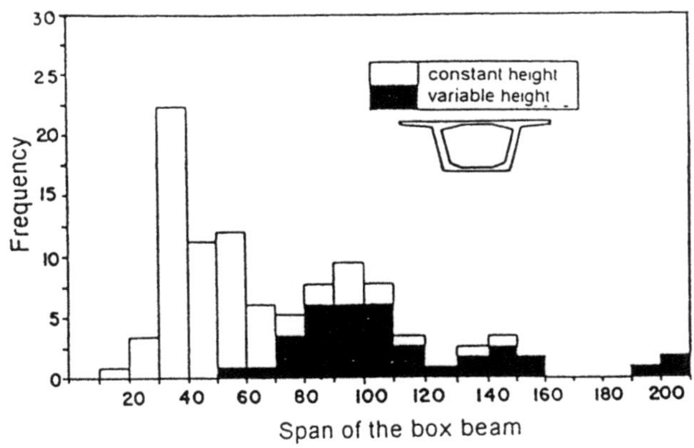


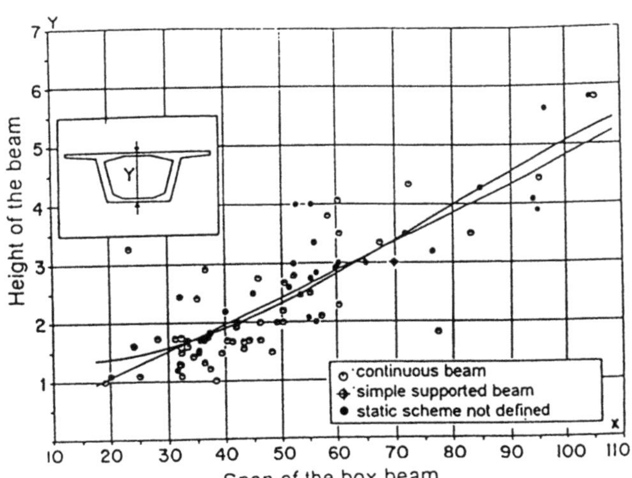


Fig. 2.15

### 2.3 Continuity systems

Figures 2.16a - 2.16b show some solutions designed to obtain the structural continuity on the piers:

1^ solution

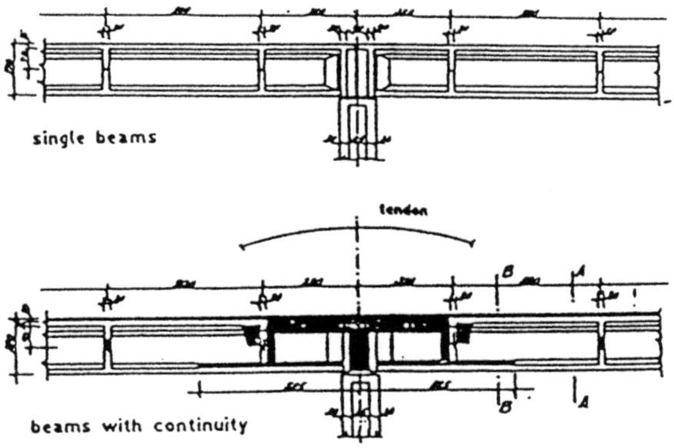


Fig. 2.16a

2^ solution

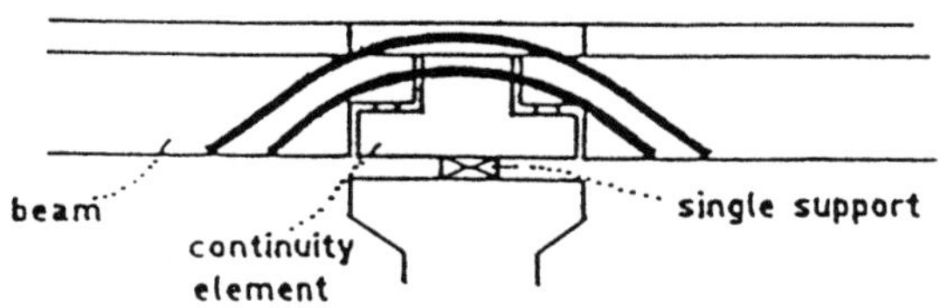


Fig. 2.16b

## 3 Structural analysis

Apart from the general problems of structural analysis it is necessary to point out some particular problems.

### 3.1 Construction phases

- It is necessary to examine the variations of the bending moments during each phase of construction method.
  Fig. 3.1 shows the basic difference that we have with the different construction methods.
- It is also important to examine the live load due to the launching steel girder during its motion and during the bridge construction.

### 3.2 Profile of prestressing cable

The profile of prestressing cable must follow the variation of bending moments from positive to negative values.

The losses of friction can be very big.

The layout of the cables, moreover, must agree with the construction method.

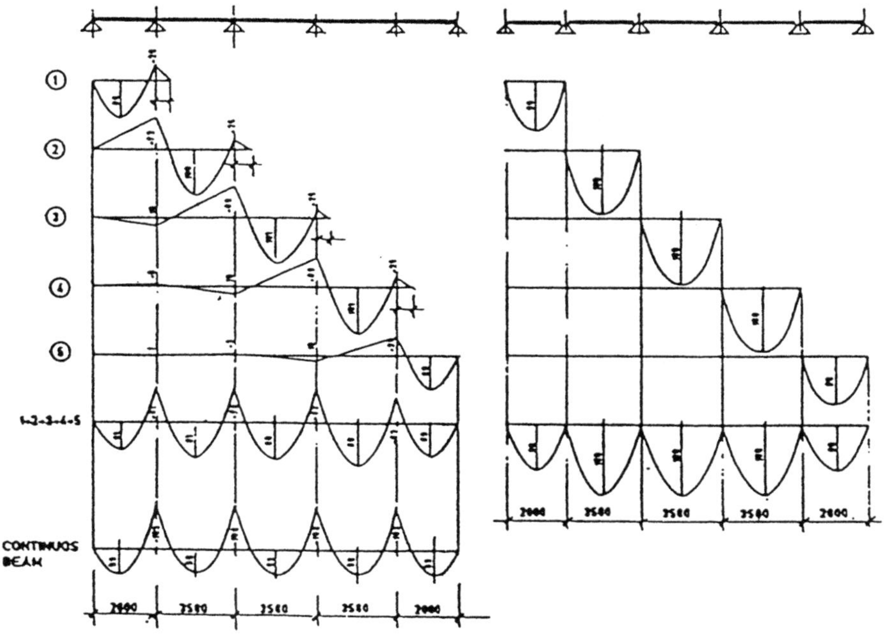


Fig. 3.1

It is convenient to use short cables for the negative bending moments. The positions of the anchorages must be studied with great attention.

**3.3 Hyperstatic actions due to tendons**
It is necessary to know the hyperstatic bending moments due to the actions of the tendons and the variations compared to isostatic bending moments.
The hyperstatic moments can be calculated by means of influence lines (Fig. 3.2)

**3.4 Wind actions**
It is necessary to take into consideration the wind's actions during the construction phases.

**3.5 Deformability of the cross-section**
The absence of the transversal connections between the ribs and the limited values of the thickness of the various decks involves some problems because the cross-section may lose its shape.
The studies can be developed with various methods of structural analysis:
Finite Element Method
Finite Strip METHOD
FOLDED PLATE THEORY
BEAM ELASTIC FONDATION ANALOGY

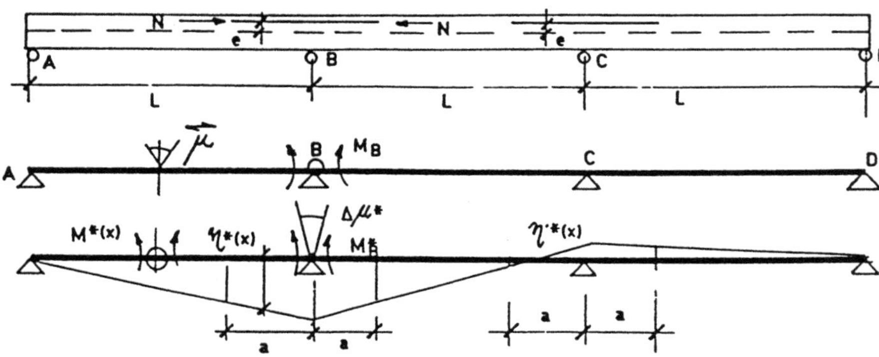


$$M_B \cdot \Delta\mu^* = M^*(x) \cdot (\mu = 1) \quad M_B = \frac{M^*(x)}{\Delta\mu^*}$$

$$M^*_B = \int_0^{2a} \eta(x) \frac{N \cdot dx}{EI} + \int_0^{2a} \eta'(x) \frac{N \cdot dx}{EI}$$

Fig. 3.2

## 3.6 Precast elements of a bridge with longitudinal linear connection

The study of these bridges is developed with a particular procedure. In Fig. 3.3 is shown the analysis scheme, with the mutual interactions. In Fig. 3.4 the congruence integral equations are reported.

With Fourier's series method, it is possible to transform this equation system in to a simple system of algebraic equations.

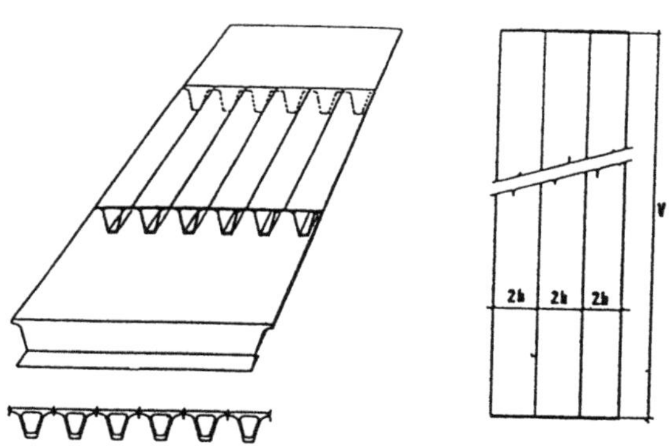

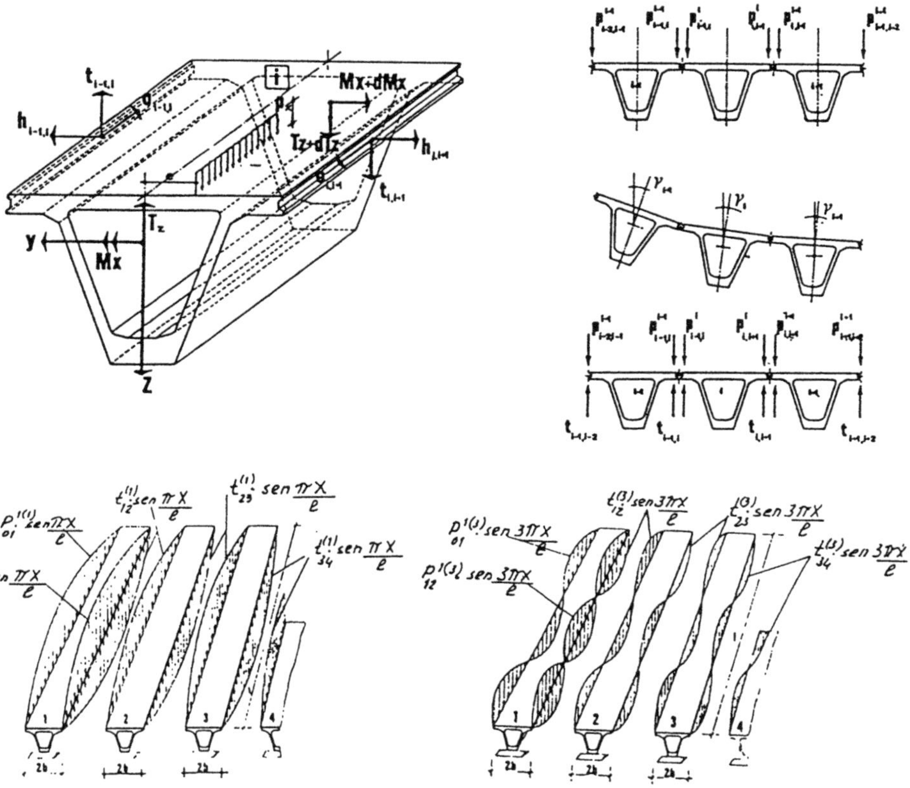


Fig. 3.3

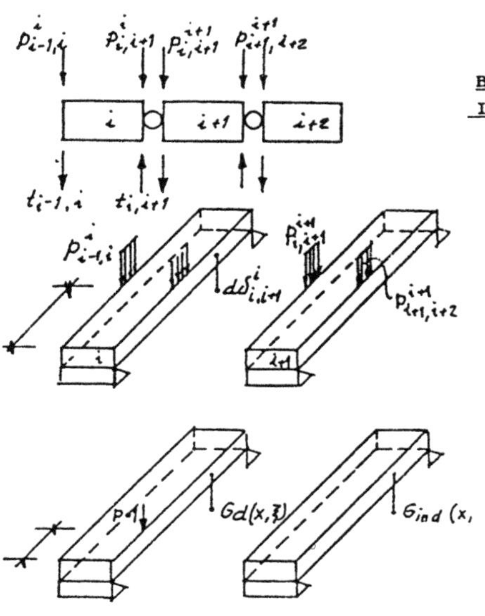

BRIDGES WITH LONGITUDINAL CONNECTIONS: INTEGRAL EQUATIONS

$$d\delta^{i}_{i,i+1} = G_d(x,\xi)\left[-t_{i,i+1}(\xi) + P^{i}_{i,i+1}(\xi)\right]d\xi$$

$$+\ G_{ind}(x,\xi)\left[+\ t_{i-1,i}(\xi) + P^{i}_{i-1,i}(\xi)\right]d\xi$$

$$d\delta^{i+1}_{i,i+1} = G_d(x,\xi)\left[-t_{i,i+1}(\xi) + P^{i+1}_{i,i+1}(\xi)\right]d\xi\ +$$

$$+\ G_{ind}(x,\xi)\left[+\ t_{i+1,i+2}(\xi) + P^{i+1}_{i+1,i,i+2}(\xi)\right]d\xi$$

$$\boxed{\delta^{i}_{i,i+1} = \delta^{i+1}_{i,i+1}}$$

$$\int_0^L G_{ind}(x,\xi)\, t_{i-1,i}(\xi)\, d\xi - 2\int_0^L G_d(x,\xi)\, t_{i,i+1}(\xi)\, d\xi + \int_0^L G_{ind}(x,\xi)\, t_{i+1,i+2}(\xi)\, d\xi =$$

$$= \int_0^L G_d(x,\xi)\left[-P^{i}_{i,i+1}(\xi) + P^{i+1}_{i+1,i+2}(\xi)\right] d\xi + G_{ind}(x,\xi)\left[-P^{i}_{i-1,i}(\xi) + P^{i+1}_{i+1,i+2}(\xi)\right] d\xi$$

SVILUPPO IN SERIE

$$P = \sum_{1}^{n} k\; p^{(k)} \operatorname{sen}\frac{k\pi\xi}{L}; \quad t(\xi) = \sum_{1}^{n} k\; t^{(k)} \operatorname{sen}\frac{k\pi\xi}{L}$$

$$t^{(k)}_{i-1,i}\int_0^L G_{ind}(x,\xi) \operatorname{sen}\frac{k\pi\xi}{L} d\xi - 2\, t^{k}_{i,i+1}\int_0^L G_d(x,\xi)\operatorname{sen}\frac{k\pi\xi}{L} d\xi + t^{(k)}_{i+1,i+2}\int_0^L G_{ind}(x,\xi)\frac{k\pi\xi}{L} d\xi =$$

$$= \left[-P^{i(k)}_{i,i+1} + P^{i+1(k)}_{i,i+1}\right]\int_0^L G_d(x,\xi)\operatorname{sen}\frac{k\pi\xi}{L} d\xi + \left[-P^{i(k)}_{i-1,i} + P^{i+1(k)}_{i+1,i+2}\right]\int_0^L G_{ind}(x,\xi)\operatorname{sen}\frac{k\pi\xi}{L} d\xi$$

$$F(x) = \int_0^L G_d(x,\xi)\operatorname{sen}\frac{k\pi\xi}{L} d\xi; \qquad a^{(k)} = \int_0^L G_d\left(\frac{L}{2},\xi\right)\operatorname{sen}\frac{k\pi\xi}{L} d\xi$$

$$F'(x) = \int_0^L G_{ind}(x,\xi)\operatorname{sen}\frac{k\pi\xi}{L} d\xi; \qquad b^{(k)} = \int_0^L G_{ind}\left(\frac{L}{2},\xi\right)\operatorname{sen}\frac{k\pi\xi}{L} d\xi$$

$$b^{(k)} t^{(k)}_{i-1,i} - 2\, a^{(k)} t^{k}_{i,i+1} + b^{(k)} t^{(k)}_{i+1,i+2} = \left[-P^{i(k)}_{i,i+1} + P^{i+1(k)}_{i,i+1}\right] \cdot a^{(k)} - P^{i(k)}_{i-1,i} + P^{i+1(k)}_{i+1,i+2}\Big] \cdot b^{(k)}$$

Fig. 3.4

### 3.7 Limit states

In each phase it is important to take into account:

- serviceability limit state
- cracking limit state

### 3.8 Curved bridges

For curved bridges all the previous problems assume different and particular complexity.

Fig. 3.5 shows the simple case of a simple beam with linear tendons.

### 3.9 Dynamic problems

It is very important the study of dynamic characteristics of the bridge.

In a seismic zones it is necessary to develop the dynamic linear and non linear analysis.

### 3.10 Bearings and expansion joints

One aspect of special interest in the structural analysis, is the dimensioning of the seats of the bearings and the expansion joints in viaducts extending over a length of more than 400 m.

In a seismic zones, in Italy, antiseismic bearing devices between the structure and the piers are studied in order to carry out the following functions:

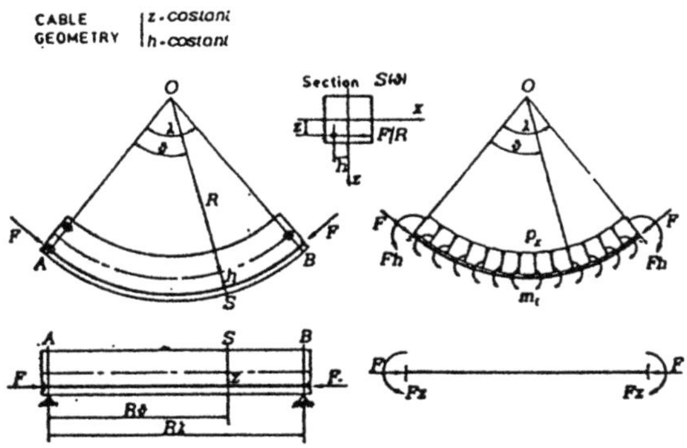


| EQUIVALENT LOADS | STRESSES |
|---|---|
| Concentrate Actions | |
| $N_A = N_B = -F$ | $N(\vartheta) = -F$ |
| $M_{xA} = M_{xB} = -Fz$ | $M_x(\vartheta) = -Fz$ |
| $M_{zA} = M_{zB} = -Fh$ | $M_z(\vartheta) = -Fh$ |
| Distributed Actions | |
| $p_z(\vartheta) = F/R$ | |
| $m_t(\vartheta) = Fz/R$ | $T_z(\vartheta) = T_x(\vartheta) = M_t(\vartheta) = 0$ |

Fig. 3.5

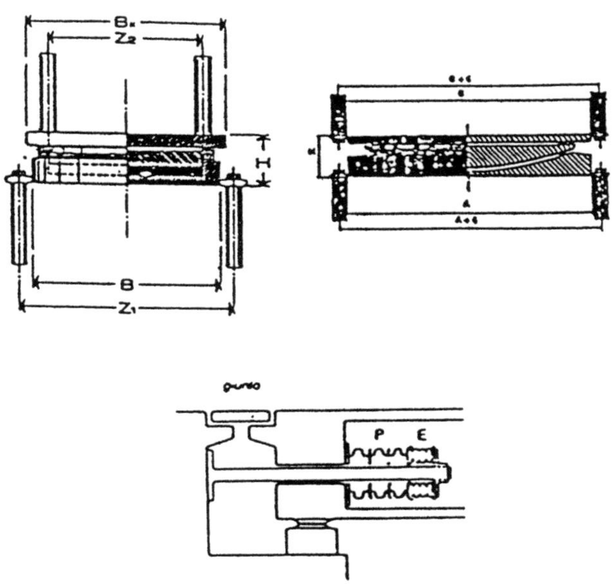


Fig. 3.6a

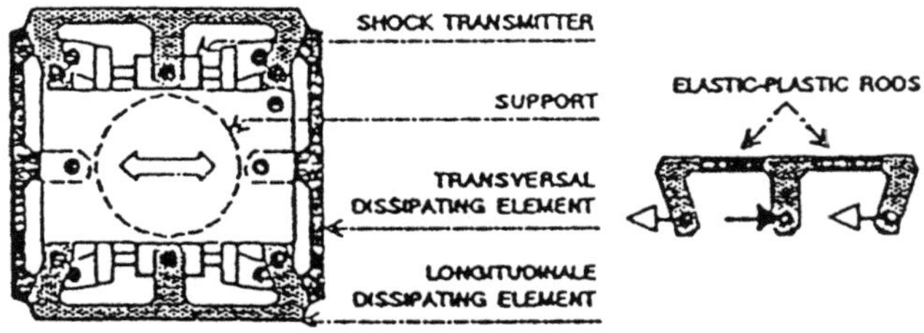


Scheme of the frame-type device.

Fig. 3.6b

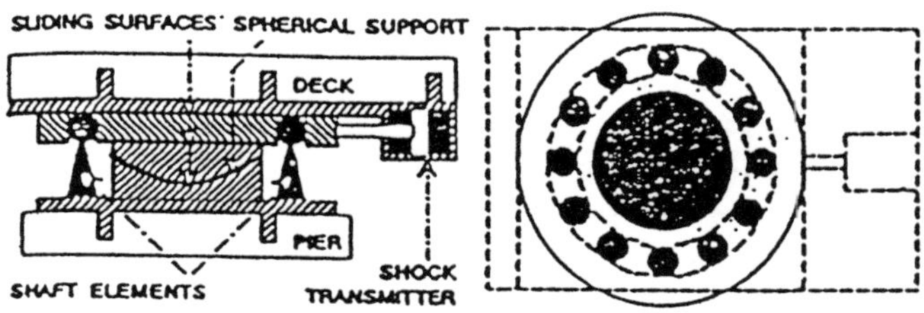


Scheme of the shaft-type elastic-plastic device.

Fig. 3.6c

- to transmit the vertical loads due to permanent and live actions
- to allow rotation
- to allow the excursion of the superstructure due to creep, shrinkage and thermal effects
- to transmit the horizontal actions to a pier without appreciable movements
- to distribute the dynamic horizontal forces (braking, design earthquake) equally among all the piers, without appreciable movements.

The expansion joint must allow the total excursion of the bridge due to creep, shrinkage, and thermal effects.
Fig. 3.6 shows a multidirectional and aseismic device.

## 4 Construction systems

The system of construction adopts advanced mechanical equipment.

These equipments provide for all cast in place phases and all assembly phases.

They are, really, mobile yards.

In the illustration of some bridges we will show the systems of equipment used for the cast in place or for the launch of beams of the bridges.

## 5 Monitoring system and maintenance

Some of the more important viaducts are provided with special monitoring equipments.
Generally the instruments adopted are the following:
- hydraulic longitudinal and transversal level meter
- probes to measure the temperatures
- biaxial inclinometers
- bearings with dynamometer for the valuation of the loads. The bearings can have a variable attitude. For example they can balance the effects of the soft ground
- transducer for the displacements
- bonded strain gage for concrete and reinforcement
- bi-axial and tri-axial accelerometers (seismic)
- traffic recorder
- transducer for meteorological events

The information is stored with event recorder

Fig. 3.7 shows the monitoring system.

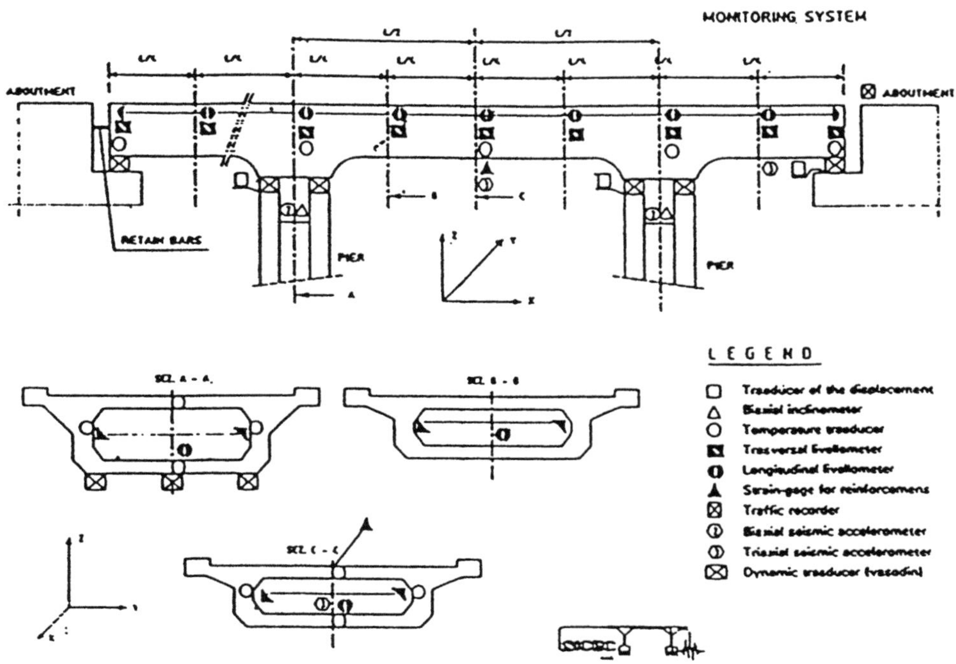


Fig. 3.7

## 6 Conclusions

In Italy, in the last years, a large number of continuous bridges have been built. A complete list would be too long for inclusion in this paper.
For this reason I am describing just a few of them.

- GRANUDA AND CASELLE VIADUCTS (MOTORWAY CARNIA TARVISIO)

- RESTELLO VIADUCT (MOTORWAY VITTORIO VENETO - PIAN DI VEDOIA)

## 7 References

AISCAT - C.N.R. (1990) Norme di manutenzione per ponti stradali.

Campo, G. Corona, G. and Mancini, G. (1990) The Torino Frejus highway bridges. Development in **Short and Medium Span Bridge Engineering.**

Capra, R. and Martinez y Cabrera, F. (1993) Il Ponte di Tadasuni. **Giornate A.I.C.A.P.**, Pisa

Dezi, L. Menditto, G. and Rosignoli, M. (1982) In tema di predimensionamento dei ponti a cassone. **L'Industria Italiana del Cemento**, n° 6.

Ghinassi, G. and Grillo, F. (1993) Analisi sismica di un viadotto in c.a.p. a travata continua con tracciato planimetrico a doppia curvatura. **Giornate A.I.C.A.P.**

Marioni, A. Marchetti, D. and Cozzani, G. (1991) La Costruzione del Viadotto Restello. **L'Edilizia**, n° 6.

Marioni, A. (1991) Antiseismic bearing devices on the Mortaiolo Viaducts. **Third World Congress on Joint, Sealing and Bearing System for Concrete structures.** Toronto, Canada. ACI.

Martinez y Cabrera, F. Rinelli, A. and Zuccolo, G. (1987) I Viadotti Granuda e Casello per l'autostrada Carnia-Tarvisio. **L'Industria Italiana del Cemento.** n° 609.

Parducci, A. (1987) Apparecchiature dissipative speciali per la riduzione della risposta sismica. **Giornate A.I.C.A.P.**

Rapino, F. Caloisi, G. and Merletti, T. (1991) Il Viadotto Restello. **Giornate A.I.C.A.P.**

**VIADUCT "GRANUDA" - "VIADUCT "CASELLO"***
**MOTORWAY CARNIA-TARVISIO (ITALY)**

## THE VIADUCT'S GENERAL DIMENSIONS

The Granuda viaduct is 320 m long overall and has eight 40,00 m spans.
The Casello viaduct is 420 m long overall, has eight 45,00 m spans, as well as two 30,00 m approach spans.
The structures have a constant width of 13,32 m, for a single carriageway(Fig. 1).
The resulting cross section that resulted is a three-cell caisson with a rectangular center cell; the side cells, incorporating the usual cantilever, are triangular, their side walls being inclined. Such a section is advisable for structures located in areas subject to low temperatures, since the deck's upper slab exposure to atmospheric agents is then transversely uniform.
The bearing and aseismic devices are those commonly known, used for all the motorway viaducts.

## THE LONGITUDINAL SEGMENTING OF THE VIADUCT DESKS

The longitudinal segmenting system used to build the decks of the Granuda and Casello viaducts presents some interesting new basic features, since it was extended for the first time here to large-scale box elements.
The technology adopted fits within the framework of the segmental precasting process, but in fact calls for the casting of span-long base elements, assembled longitudinally. This process, in many ways a counterpart of the transverse segmenting system, marks a noteworthy advance in viaduct-construction methods, having a number of statics and construction advantages.

## THE CONSTRUCTION PHASES OF THE DECKS

Fundamentally, the construction process broke down into the segmenting of the caisson deck girder converting the whole span into two halves by passing a vertical plane through its longitudinal axis, and into the connection of the half-caissons by two continuous linear joints. Fig. 2.

* FEDERICI COMPANY - (ROME) - S.A.L.C. COMPANY (PADOVA)
Designer: F. Martinez y Cabrera (Milano) - A. Rinelli (Padova) - G. Zuccolo (Padova)

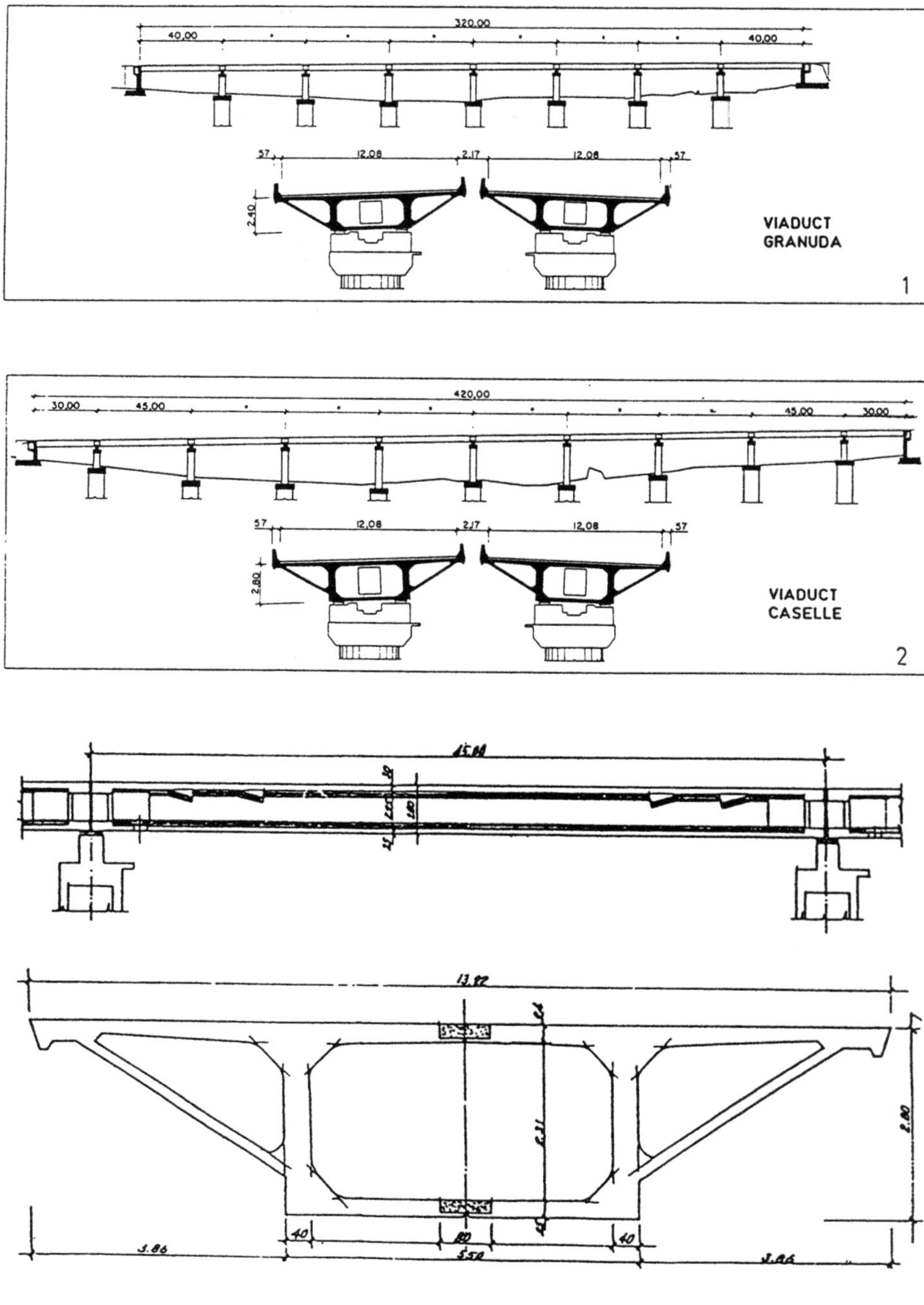
320,00
40,00
40,00
57
12,08
2,17
12,08
57
2,40
VIADUCT
GRANUDA
1
420,00
30,00
45,00
45,00
30,00
57
12,08
2,17
12,08
57
2,80
VIADUCT
CASELLE
2
13,82
2,80
3,86
40
80
40
5,50
3,86

Fig. 1

The decks were built in the following phases:

a) precasting at the plant of each half-caisson over its whole length, its steam curing, and the other phases preliminary to prestressing to prepare it for haul;

b) hoisting by hydraulic devices, hauling on rails and laying in position using high-capacity launch equipment; the precast elements are put side by side in place on special complex bases set up to adjust their position in line and grade;

c) casting the connection for top and bottom slabs, along the interface lines;

d) laying the continuity cables between the launched span elements, execution of the connection casts between the girder ends, and tensioning of some continuity cables;

e) completion of cable tensioning operations when the successive spans had been launched;

f) finish works: curbs and paving.

This solution was advantageous in several ways:

- half-caissons poured in a few hours in a stable and protected yard, with full control of materials and of the curing process. This meant true longitudinal element monolithicity, difficult to obtain using other procedures. Thus, the longitudinal cross section is kept integral, with no statics interruptions due to bending and shear;
- none of the interface problems inherent to other structural systems;
- the structure as a whole is more monolithic than in-situ-poured caissons. These latter are necessarily built *"in installments"*: a first cast of the bottom, slab, then the cast of the walls, and finally the top slab. This means a larger number of successive castright in the statically-engaged areas, that is, in the nodes of the caisson cross section;
- the longitudinal segmenting involves only problems tied to local behaviour and, as regards the overall statics, tied to its torsional behaviour in response to antisymmetric loading conditions;
- the effects due to deferred elastic phenomena are more easily controlled, because of the precast structural element longitudinal homogeneity.

  The continuous construction cycle described meant, using two outside forms, an average construction time of three days for each half-caisson (corresponding to 300 mq of deck area); after this the launch, assembly and joining phases these not slowing down the advance of the bridge.

a) Static analysis

Numerical analysis was used to study the caisson, using a finite-strip discretizing procedure, and treating it as a plate-slab. This method can be used when the elastic and

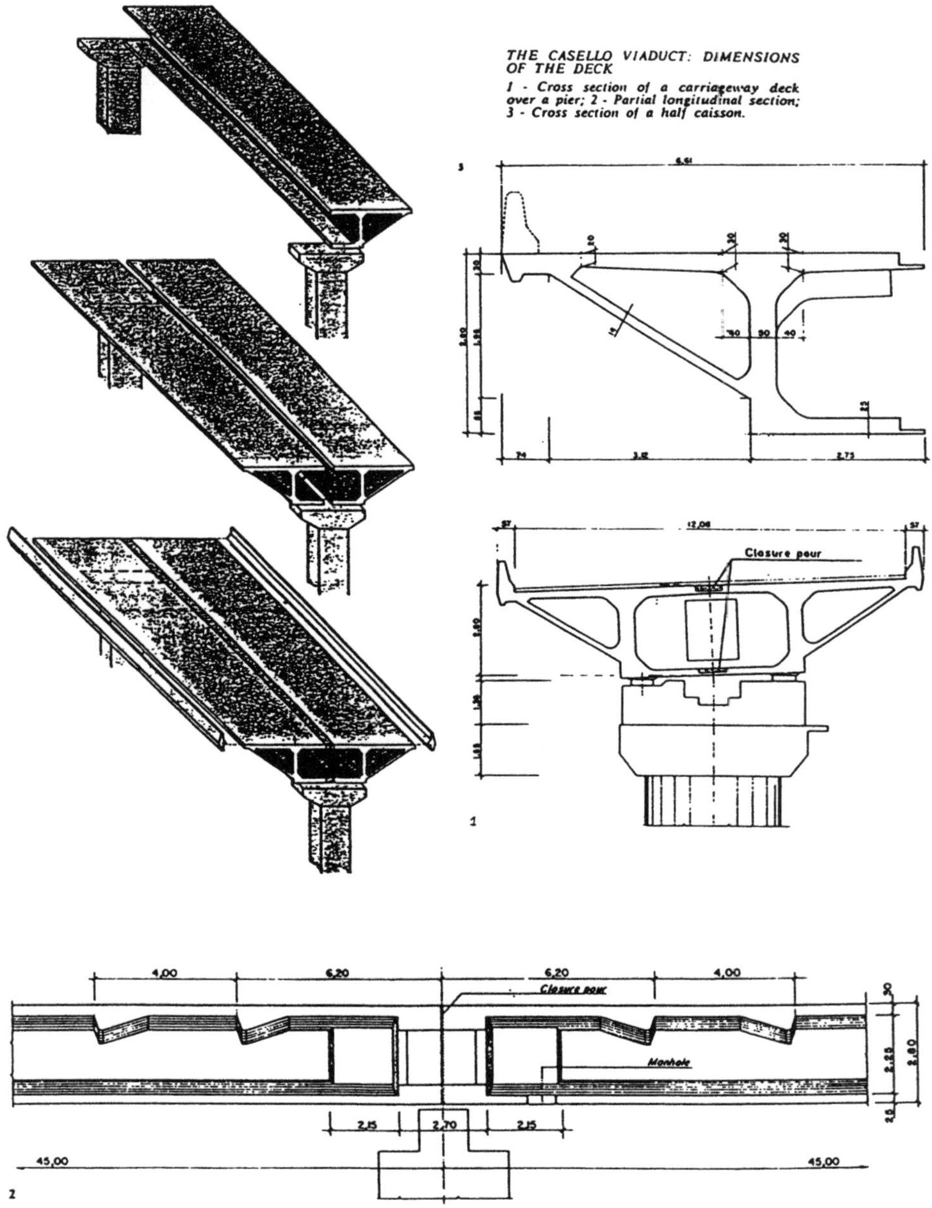
THE CASELLO VIADUCT: DIMENSIONS OF THE DECK
1 - Cross section of a carriageway deck over a pier; 2 - Partial longitudinal section; 3 - Cross section of a half caisson.
3
1
Closure pour
4,00
6,20
6,20
4,00
Closure pour
Manhole
2,15
2,70
2,15
45,00
45,00
2

Fig. 2

geometric parameters stay constant in the longitudinal direction, only the amount of the load varying. (Fig. 3).

b) Experimental analysis

On the basis of the above analysis it can be stated that in longitudinal assembly systems the connection joints, even though of considerable importance, are not in themselves a critical factor for the structure's safety.

Nonetheless experimental investigations were undertaken to study the joints real behaviour and their aptness to bring about the desired cross-wise static continuity.

Full scale models were then constructed of the area of the slab affected by the joint (Fig. 4), reproducing a 1,45 m-long strip from the caisson, with a transversal profile similar to the central area of the caisson, including the parts where the side walls were attached. Fig. 4 shows the model dimension, together with some of the loading conditions studied:

- two uniform strip loads of dimensions 10 x 145, astride the transversal mid-span and spaced 1,20 m apart;
- two load "imprints" of dimensions 65 x 25, spaced at 1,80 m;
- one isolated load "imprint" of dimensions 65 x 25, placed at the center of the slab.

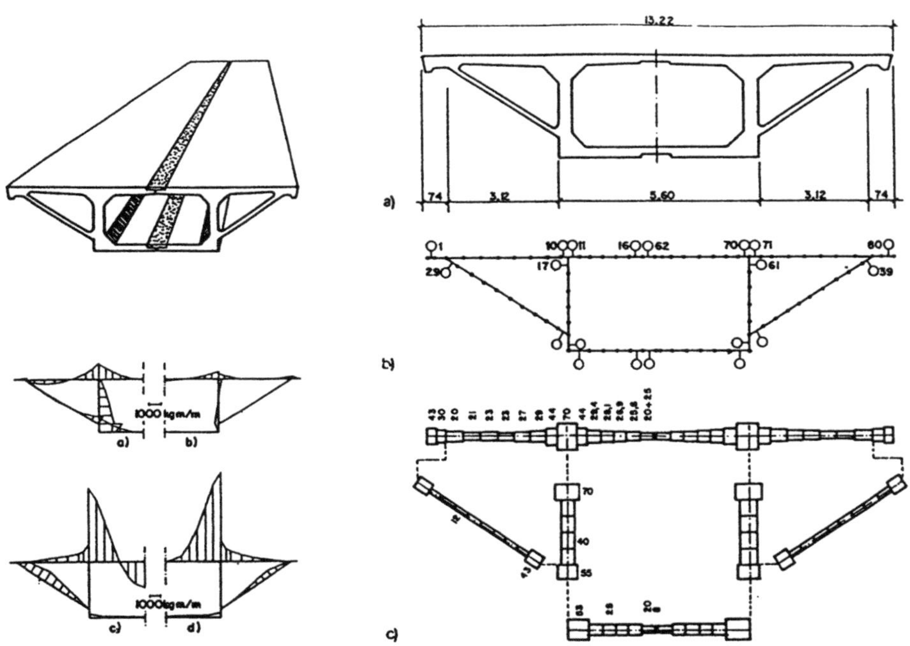


Fig. 3

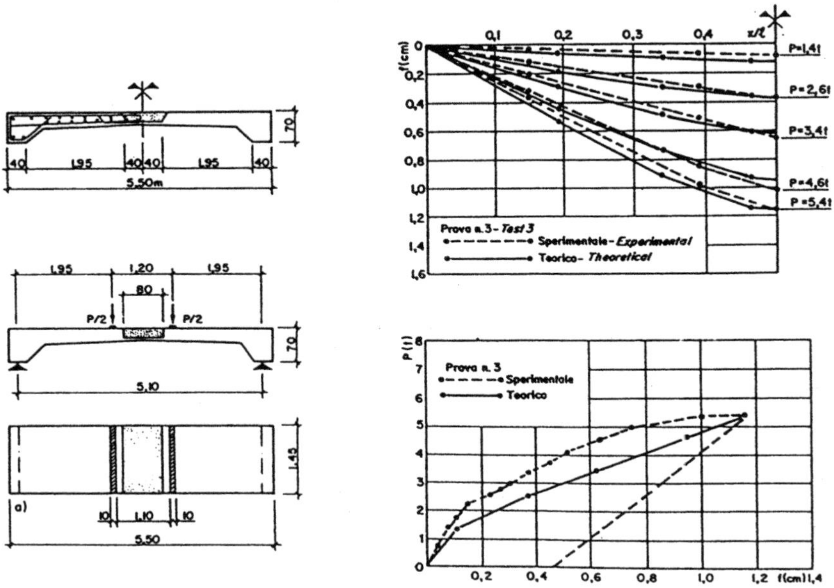
5,50m
1,95
1,20
80
P/2
5,10
a)
5,50
Prova n. 3 - Test 3
Sperimentale - Experimental
Teorico - Theoretical
P=1,4t
P=2,6t
P=3,4t
P=4,6t
P=5,4t
f(cm)
x/ℓ
Prova n. 3
Sperimentale
Teorico
P (t)

Fig. 4

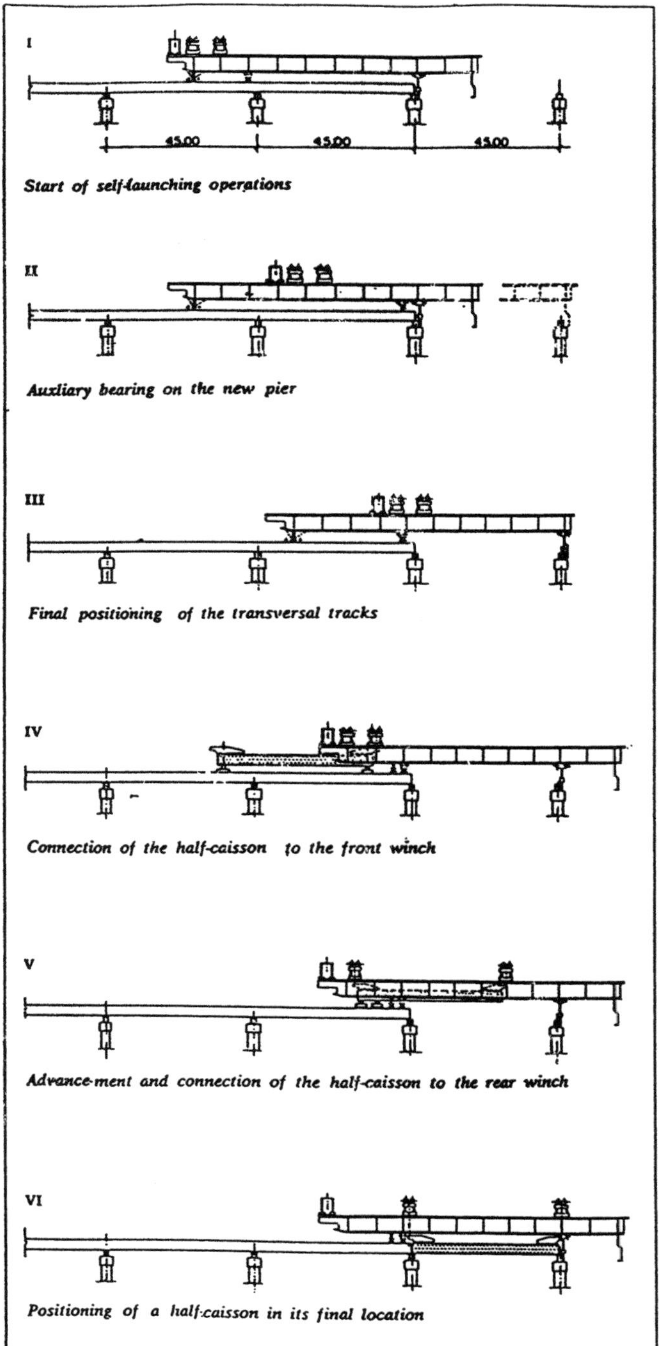
I
45.00
45.00
45.00
Start of self-launching operations
II
Auxliary bearing on the new pier
III
Final positioning of the transversal tracks
IV
Connection of the half-caisson to the front winch
V
Advance-ment and connection of the half-caisson to the rear winch
VI
Positioning of a half-caisson in its final location

Fig. 5

MOTORWAY OF FREJUS (ITALY)

CLAREA AND RAMAT VIADUCTS*

| | | |
|---|---|---|
| TOTAL LENGHT | : | CLAREA $L_{TOT}$ = 600 m<br>RAMAT $L_{TOT}$ = 900 m |
| SINGLE SPAN | : | l = 100 m |
| CROSS SECTION | : | monocellular box girder 10.55 m wide and variable height from 2,45 m to 4,55 m with parabolic law (two carriageway) |
| PIERS | : | box type up to 92 m in height |
| SCHEME | : | continuos beam on seven spans |
| TYPE | : | segmental bridge built assembling precast segments by mean of post tensioned tendons (trasversal segmentation).<br>In the assembling procedure a steel launching girder has been adopted, having a total weight of 280 tons.<br>The launching girder was capable of handling segments up to 120 t in weight. |
| BRIDGE TO PIER CONTACT | : | the contact is obtained by means of two lines of three bearings each, spaced 3,90 m apart (reduction of 10% in the positive bending moment). |
| SPEED OF COSTRUCTION | : | costruction of two complete half span in 21 days (4,76 m/day). |
| MATERIALS | : | concrete 0,78 mc/$m^2$<br>ordinary reinforcement 110 kg/$m^2$<br>presstressed reinforcement 45 kg/$m^2$ |

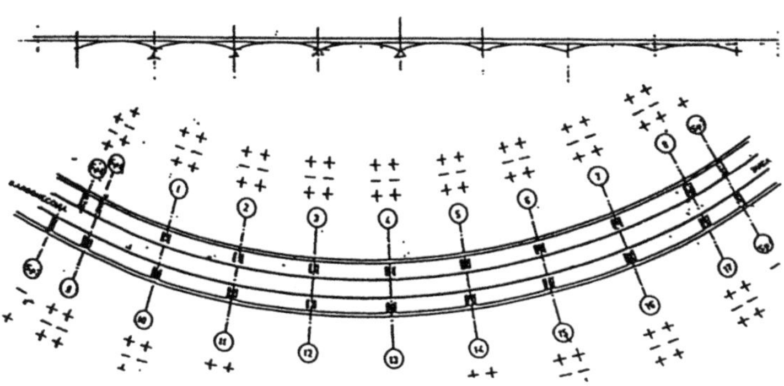

---

* RIZZANI DE ECKER
  Designer: G. Corona (Turin) - G. Mancini (Turin) - P. Napoli (Turin)

**CLAREA VIADUCT ( Frejus motorway - Italy )**

Contractor: RIZZANI DE ECKER
Designer: Ing. G. Corona
Ing. G. Mancini
Ing. P. Napoli

**MOTORWAY VITTORIO VENETO - PIAN DI VEDOIA**

**RESTELLO VIADUCT**

| | | |
|---|---|---|
| TOTAL LENGHT | : | $L_{TOT}$ = 2100 m |
| SINGLE SPAN | : | l = 48 m; 72 m; 60 m; 100 m |
| CROSS SECTION | : | monocellular box girder 12,88 m wide and, variable height from 3,70 m to 5,00 m with linear law (two carriageway). |
| PIERS | : | box type up to 60 m in height. |
| SCHEME | : | continuos beam |
| TYPE | : | segmental bridge built assembling large precast segments by mean of post-tensioned tendons (trasversal segmentation).<br>The segments are of two basic type: one segment with variable section and 40 m lenght and another type with constant section and various lenght (32 m; 40 m; 60 m)<br>In the assembling procedure a steel launching girder has been adopted.<br>The launching girder was capable of handling segments up to 800 t in weight. |
| BRIDGE TO PIER CONTACT | : | two multidirectional bearing with longitudinal excursion $S_L$ = 1000 mm and trasversal excursion $S_T$ = 15 mm.<br>The viaduct has aseismic longitudinal devices at the aboutments and transverse devices on the piers. |
| MATERIALS | : | concrete 0,90 mc/m$^2$<br>ordinary reinforcement 100 kg/m$^2$<br>presstressed reinforcement 50 kg/m$^2$ |

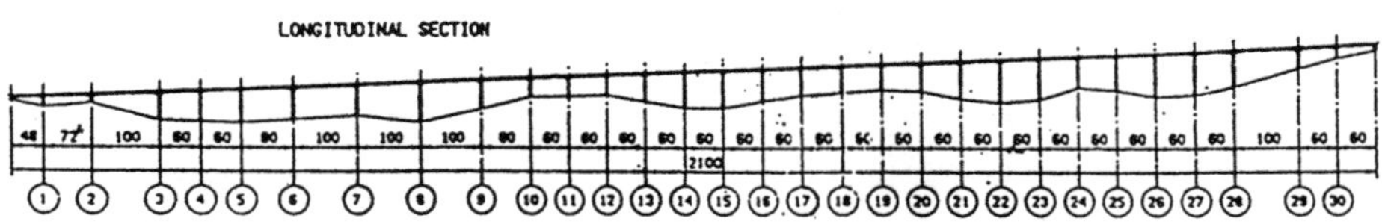


Contractor: GICO COMPANY
Designer: Italprogetti S.r.l. - Roma - Ing. Caloisi
Engineer's Representative: S.P.E.A . - Milan - Ing. T. Merletti

Autostrada Vittorio Veneto Pian di Vedoia (Italy)
Viadotto Restello (continous beam L=2100m)

A. Spans type

A

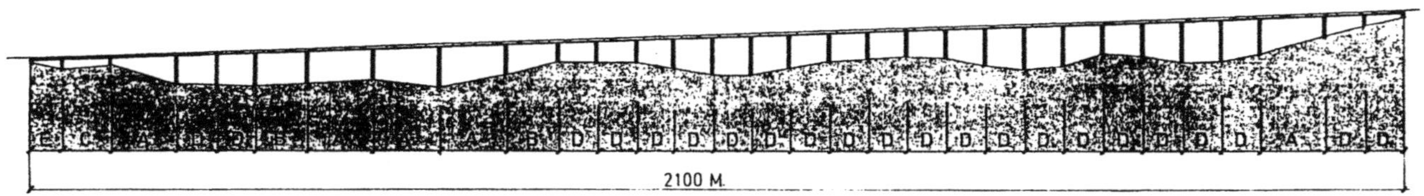


(A = 100 m; B = 80 m; C = 72 m; D = 60 m; E = 48 m; )

B

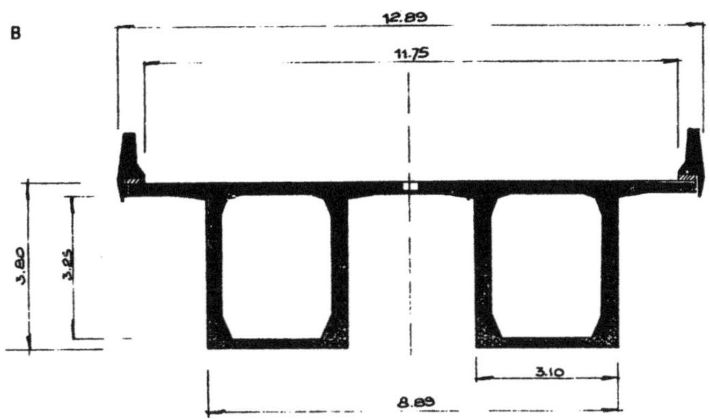


B. Section at span center

C

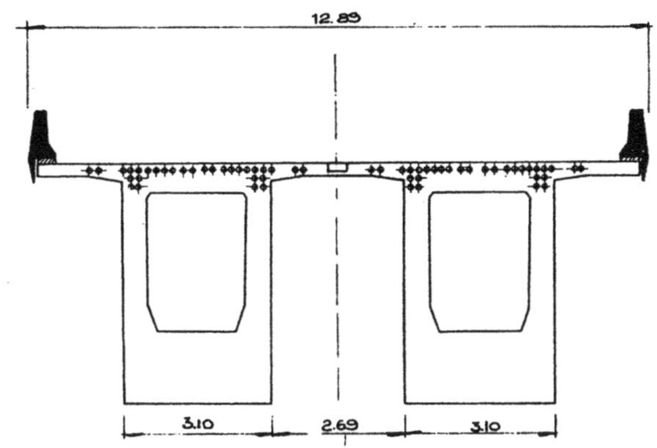


/ C. Section at the pier

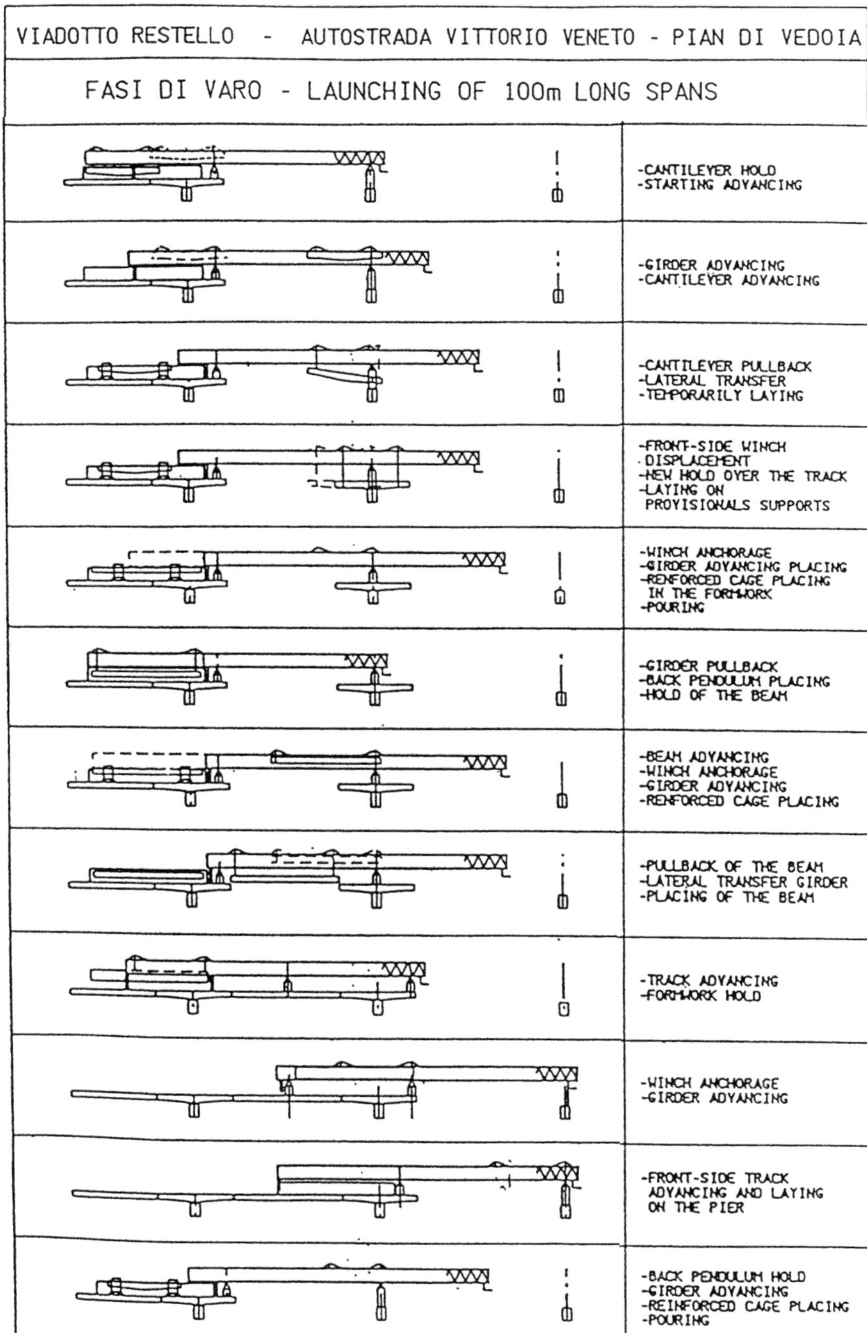
VIADOTTO RESTELLO - AUTOSTRADA VITTORIO VENETO - PIAN DI VEDOIA
FASI DI VARO - LAUNCHING OF 100m LONG SPANS
-CANTILEVER HOLD
-STARTING ADVANCING
-GIRDER ADVANCING
-CANTILEVER ADVANCING
-CANTILEVER PULLBACK
-LATERAL TRANSFER
-TEMPORARILY LAYING
-FRONT-SIDE WINCH
DISPLACEMENT
-NEW HOLD OVER THE TRACK
-LAYING ON
PROVISIONALS SUPPORTS
-WINCH ANCHORAGE
-GIRDER ADVANCING PLACING
-RENFORCED CAGE PLACING
IN THE FORMWORK
-POURING
-GIRDER PULLBACK
-BACK PENDULUM PLACING
-HOLD OF THE BEAM
-BEAM ADVANCING
-WINCH ANCHORAGE
-GIRDER ADVANCING
-RENFORCED CAGE PLACING
-PULLBACK OF THE BEAM
-LATERAL TRANSFER GIRDER
-PLACING OF THE BEAM
-TRACK ADVANCING
-FORMWORK HOLD
-WINCH ANCHORAGE
-GIRDER ADVANCING
-FRONT-SIDE TRACK
ADVANCING AND LAYING
ON THE PIER
-BACK PENDULUM HOLD
-GIRDER ADVANCING
-REINFORCED CAGE PLACING
-POURING

General view of the work in progress

The mobile formwork is always just behind the launching equipment

A precast girder ready to be launched

60 m girder launching

View of the front rail beam

Launching of the top pier girder

Operator at control panel

# 13 THE UK DEPARTMENT OF TRANSPORT VIEW ON CONTINUITY/INTEGRAL BRIDGES

G.P. PATERSON
Department of Transport, London, UK

Abstract
The Department of Transport will shortly publish a new Departmental Standard and Advice Note on "Design for Durability", whose requirements will apply to all new bridges built for the Department. The need to make new bridges continuous over intermediate piers, and, for short and medium spans, fully integral at their ends, is an important part of these new requirements. The paper describes the development of the documents and gives details of the requirements for continuous and integral bridges, together with some explanation of the thinking which lay behind the development of these documents.
Keywords: Bridge, Durability, Standards, Continuous, Buried structures.

## 1 Introduction

The Department of Transport has been aware, at least since the mid 1970s, that the performance of its stock of bridges built in the post-World War II years was not as satisfactory as had been expected. One of the main causes of this was the deterioration caused to the ends of bridge decks and to substructure elements below them due to the presence of full joints at the ends of the decks, and at the ends of each span in multi-span bridges. Accordingly, individual submissions for approval in principle started to be vetted with a view to eliminating fully discontinuous joints at intermediate piers.

The "Maunsell Report", (1), published in 1989, was a survey of a randomly-chosen set of 200 bridges, and its findings corroborated, among other things, the perception that the presence of full deck joints was a major source of deterioration in the bridge stock as a whole. By the time this report was published, however, the Department was already preparing a first draft of an advice note on Design for Durability, in which the issue of bridge continuity was to be of central importance.

## 2 Development of the Department's Draft Advice Note

In its early draft stages, the Advice Note made a strong recommendation that full or partial continuity should be used at all intermediate piers, and that half-joints be totally avoided wherever possible. It also continued with this logic to recommend the use of fully-integral deck ends, that is to say, the use of portal-type bridges with the abutment stems integral with the ends of the deck. However, received good practice was that such bridge types could only be used for quite short bridge lengths, and therefore portals were generally thought suitable only for single spans of 10m or less.

It gradually became obvious that this perception of good practice was very conservative, and that much greater bridge lengths could be constructed as integral bridges. Experience from the U.S.A. particularly from the State of Tennessee, and from Sweden, suggested that lengths of the order of 100m could be achieved without incurring problems. Since the distribution of the numbers of our stock of bridges is highly skewed towards the shorter bridge lengths, the application of a requirement that bridges shorter than a length of, say, 100m should be designed as integral wherever possible would mean that the overwhelming majority of all our new bridges would be of the integral type.

The limitation on the length of integral bridges is imposed by the response of the soil mass behind the abutments. Long integral bridges experience large changes in length as a result of thermal effects; in warm periods, their ends push against the soil mass and force it to deform in a compressive manner; given the high relative strength and stiffness of structural materials, the movements of expanding decks can be considered as almost irresistible, although stiff soils may induce significant compressive stresses in the deck as well.

In cold periods, the deck shrinks and, depending on its nature, the soil will either expand to wholly fill the resulting void, or it may relax to a lesser degree, leaving a void at the interface until the next expansive cycle closes the gap. It is the cyclic nature of these movements which causes concern to bridge designers. Both daily and seasonal cycles are involved, and the number of cycles is large in the life-time of a typical bridge. It is difficult to foresee the cumulative effects of many such cycles; perhaps the soil's resistance builds up progressively after many cycles, implying a gradual vertical compaction of the soil mass due to cumulative secondary effects; or perhaps some soils have the ability to cope with deck movements by creeping away from an expanding deck and towards a contracting one. The Department has commissioned research through TRL to explore these issues in more depth. Meantime, existing findings such as those of Broms and Ingelson (2) may be used for design.

However, this may be, experience abroad was suggesting that, in practice, no serious problems were arising in the management of such long integral bridges. It was decided to check the Department's bridges data-base to see what were the longest integral bridges in our own inventory, and what, if any, problems had been experienced with them. Given the widespread acceptance of the upper limit of about 10m as traditional good practice, we were surprised to find that the English Trunk Road Network contains a large number of integral bridges much longer than this. There are at least 20 bridges with lengths in the 35-50m range; the longest bridge we have identified is about 110m long, although its skew of about 60° corresponds to a right span of approximately 60m. Many of these bridges had been built in the 1960's and 1970's as part of the development of the motorway system, and had been performing in a remarkably trouble-free way in service, to the extent that the significance of their integral nature had been largely overlooked.

Hence it is now proposed that the Department should require that all bridges up to lengths of about 60 or 70m should be made with fully integral abutments and with full or partial continuity at intermediate piers where suitable. The final form of the advice note will reflect this, and a short Departmental Standard will give mandatory force to this and certain other requirements relating to bridge durability. The drafts of these two documents are now being piloted through their technical and policy review committee stages within the Department, and, all being well, both should be published before the end of 1993. As a result of these review procedures, some of the requirements described later in this paper may be subject to change.

## 3 Contents of the Department's Advice Note

The Advice Note recommends the use of continuous bridge decks wherever possible, even in situations where considerable relative settlements may occur between individual piers. Due allowance should be made for predicted relative movements in the design of the deck elements. Where these effects cannot be catered for using full continuity, partial continuity may be preferred, that is, the use of simply supported beams with a continuous deck slab. However, it is pointed out that, in fully-continuous decks, the ability to cope with relative settlements is enhanced by the use of higher span/depth ratios. Span/depth ratios for continuous bridges tend to lie in the range between 20 and 25, but ratios of 30 or even 35 are achievable for the same deck strength, though with a progressive loss of economy at the construction stage. The Advice Note points out that, if such slender decks are used, care needs to be taken to ensure that the decks are not too "lively".

Whereas with reinforced concrete, steel and steel/concrete composite decks full continuity will generally be provided, there is a problem with the use of pretensioned prestressed concrete beams, which do not obviously lend themselves to continuous applications. Full continuity can be achieved by various methods, but some uncertainty remains about the final distribution of dead-load stresses, due to the difficulty in predicting accurately the development in time of the effects of creep, shrinkage and other losses in the beam elements. There can be very significant redistri- bution of dead-load moments between support and mid-span areas due to this effect. Partial continuity may be preferred, though the detailing at pier heads is likely to be more clumsy, with separate sets of bearings for each span. Several alternative forms of partial continuity are available; I personally prefer the form which has been used extensively in Germany and was pioneered in UK by Dr A. Kumar, involving the use of a compressible layer between the beams and the slab, so that the deflection profiles of each under load can be entirely independent over a certain zone at the pier position.

The Advice Note goes on to recommend the use of fully-integral structures, that is, structures with full or partial continuity at piers, and with abutments (and possibly also pier-heads) built integrally with the bridge deck. Deck expansion joints would be totally eliminated by this recommendation; if pier-heads are integral, bearings would also be dispensed with entirely.

The use of run-on slabs at bridge abutments is discussed. This may have several advantages when used with integral bridges. The most important of these is to prevent the occurrence of vertical discontinuities in the final road surface, which might be caused by progressive over-compaction of the soil locally behind the end of the bridge. Such discontinuities can be hazardous to traffic, and can cause big increases in the dynamic effects of vehicle axles on bridge decks. Run-on slabs ensure that only slight rotational discontinuities occur at their ends, along with horizontal discontinuities at their outer ends only. Another advantage of the use of run-on slabs is that the soil mass immediately behind the abutment may be left only partly compacted, with possible benefits in reducing soil pressures on abutment stems; in general we do not favour the US practice of using wholly uncompacted fill below run-on slabs. Conservative guidance on the design of reinforcement for run-on slabs is given.

Obviously, the magnitude of soil pressures generated by long integral bridges can be large, reaching passive pressure values. The use of slender abutment stems may mitigate these effects to some extent, since the large excursions will then be more limited to the upper levels of the soil mass. This suggests the use of "balanced-design" for R.C. stems, that is, the concrete section is not thicker than is needed to provide a compressive couple equal to the tensile couple of the reinforcement. The Advice Note does not give guidance on the calculation of earth pressures, since it is assumed that standard theories can be used to derive these from the structural movements involved. Designers may need to assess the thrust forces communicated by the soil to the deck, and allow for this in building up the overall stresses in the deck cross-section. The general guidance on these problems is that, although substantial additional costs may be incurred in dealing with soil pressures on long integral bridges, the benefits of reduced maintenance and more durable structures are likely to far outweigh them.

Although the general treatment of integral abutment design is in terms of the use of standard full-height abutment walls, we are aware that bank-seat abutments are also very common, and that use of this type is likely to increase even further in future. Much of the discussion is common to both types, of course, but explicit reference is also made to integral bank-seats. Both U.S. and Swedish practice has developed specific ways of dealing with this case, and the former is referred to in the Adivce Note, with an illustration of the typical arrangement used.

The Advice Note sets out the various recommendations for the design of integral bridges as outlined above. The Standard requires that all bridges shall be designed as continuous over intermediate piers with fully integral abutments for bridge lengths up to 70m (the precise figure may vary slightly in the finally published version). Clearly, this is not going to be possible in every case; however, we expect the number of instances where recourse to "Departures from Standard" is required to be quite few.

In line with the promotion of integral bridge structures, the Standard also requires that, where adequate clearances exist, buried structures should be used. Apart from the obvious point that buried structures are normally integral structure types, (e.g. R.C. box culverts and corrugated-steel buried structures) which are not exposed to the problems of joint leakage, the use of all buried structure types, including the integral types, have two further advantages:-

- being relatively remote from the highway running surface, problems of aerial salt spray and of water penetration aided by the locally high hydraulic overpressures due to wheel loads are reduced in significance
- being structurally separate from the highway structure, maintenance of both is eased.

## 4 Conclusion

The coverage of the Departments' Standard and Advice Note on Design for Durability is much more extensive than the treatment of continuous and integral bridge types. For example, there is an important section on the adoption of structure types which minimize or avoid the use of tensile reinforcement in concrete structures; as part of this section, the use of the arch type of structure, which is of course another integral structure form, is advocated. The Department's participation in a number of initiatives to build arch bridges, both in concrete and in masonry, is aimed at re-establishing this bridge type as a significant option for new bridge procurement.

Nevertheless, the detailed coverage of the requirements for continuous and integral bridges, and the reinforcement of these requirements by use of a Departmental Standard is likely to lead to significant changes in the overall pattern of our procurement. This should lead to a progressive change in the balance of our bridge stock towards simpler, more easily-maintained bridges, with reduced needs for bearings and expansion joints and for the troublesome facilities needed to service them.

## 5 References

Broms, B.B. and Ingleson, I. **Earth Pressure against the Abutments of a Rigid Frame Bridge**, Géotechnique 21, No. 1.

Wallbank, E.J. (April 1989) **The Performance of Concrete in Bridges**, HMSO, London

# 14 JOINTLESS STEEL VIADUCTS FOR RAILWAY

E.C. HAMBLY
Edmund Hambly Ltd, Berkhamsted, UK
G.W. OWENS
Steel Construction Institute, Ascot, UK

Abstract
This paper examines the feasibility of using jointless, thermally restrained construction for railway viaducts. Prima facie, it demonstrates that such superstructures are technically acceptable for both strength and fatigue limit states. It also shows that the abutments and piers can be designed satisfactorily. Such construction is most likely to be economic for multispan viaducts. It is likely to show considerable economies in service; unlike conventional construction, there are no incompatibilities of thermal movement between the rails, superstructure and substructure, maintenance costs should therefore be reduced.
Key words: continuous, bridges, steel, railway, jointless, maintenance

## 1 Introduction

Steel viaducts can be designed up to any length without movement joints. Such viaducts would be particularly suitable for high speed railways. The behaviour of a jointless viaduct under thermal and live loading would be essentially the same as the behaviour of the continuous welded railway track. The absence of any expansion joints in the structure obviates the need for any joints or other special measures in the track construction. It also eliminates sources of noise at bearings and joints during passage of trains.

It has been demonstrated in North America that concrete and steel viaducts can function satisfactorily without expansion joints at either abutment for lengths of up to 200m, Burke (1990). These viaducts expand and contract during temperature changes; though in general the daily and annual cycles of temperature strain are much smaller than are calculated for code extreme ranges of temperature. The thermal movements that do occur are accommodated by deformations within the embankments behind the abutments. However, for high speed railways these small movements still cause problems because the structures change in length while the continuous railway track above does not. This is one of the reasons why little use has been made of continuous girders for railway viaducts. Steel railway bridges have generally been simply supported, and have tended to be more noisy in service than concrete (though composite construction and ballast do reduce noise considerably).

The viaduct proposed here has anchored abutments which do not move.

*Continuous and Integral Bridges.* Edited by B.P. Pritchard. Published in 1994
by Taylor & Francis, 2 Park Square, Milton Park, Abingdon, Oxon, OX14 4RN. ISBN: 0 419 19030 9.

Instead of expanding or contracting during temperature change, the steel is subjected to locked-in compression or tension stresses from end to end, in the same manner as continuous railway track. At the ends these stresses are transferred into the abutments and then into anchorage stresses in the ground. The abutments are more massive than conventional abutments, however their anchorage requirement does not increase with the length of the viaduct; so jointless construction becomes more economic as the length increases.

Longitudinal traction and braking forces are transmitted along the full length of the viaduct to the abutment anchorages. As a result relatively slender piers can be used for all intermediate supports. The viaduct could be constructed at the abutments and launched incrementally, in order to minimise the interference with the ground below. Once the superstructure has been made continuous with the abutments, the bearings used for construction can be fixed. If provision has to be made for replacement of spans in the future the abutments and bearings can be designed so that they can be unfixed while the replacement takes place.

A check has to be made of the fatigue from thermal stress cycles at welded joints and at the holes for friction grip bolts; however this fatigue does not appear to be critical as compared to the design for live loading.

It is found that the maximum stress range induced by extreme hot and cold temperatures is small compared to the stress range capacity of high strength steel between compression and tension limits. In contrast, a jointless viaduct of concrete appears to be less practicable because the stress range between extreme temperatures would use up a high proportion of the compression capacity of the material. The system is also inherently less suitable for concrete because of the much greater cross-sectional area of bridges of comparable span for that material. Even allowing for differences in modulus, the thermal forces induced in a concrete superstructure would be several times greater than in a steel superstructure.

The jointless steel viaduct demonstrated here is assumed to have conventional protective treatment, or to be of weathering steel for reduced maintenance. However, economy might be achieved if the structure was shielded against extreme temperatures with a temperature control envelope or insulation layer.

## 2 Arrangement and appearance

Figure 1 illustrates the general arrangement for the jointless steel viaduct. This arrangement differs significantly from a viaduct with joints because:-

1) Spans may be longer due to continuity over supports.
2) Piers may be more slender because braking and traction forces are transmitted to abutments.

The jointless viaduct is shown with two continuous longitudinal steel plate girders, with composite cross girders. The depth has been set at about 3.5m in order to provide a 2m height of sound barrier together with a reasonable depth of ballast.

The spans of up to 50m have been based on this depth of plate girder.

The design principles demonstrated here are applicable to viaducts with box girders.

## 3 Girder design

### 3.1 Temperature cycles

BS5400 and TRRL 696 provide temperature data primarily for highway bridges with girders located below slabs. The jointless viaduct in Figure 1 has markedly different thermal properties and considerable economy might be achieved by research into the thermal behaviour of specific structural forms with different coatings and envelopes, and by calculations based on first principles as in Hambly (1978). In order to be conservative, in the absence of specific information, this example is based on data in BS5400.

Three distinct types of temperature cycle affect the fatigue of welded joints during the 120 year design life:-

1) One extreme cycle during the 120-year design life between the extreme maximum and extreme minimum temperatures.
2) 120 annual cycles between summer and winter temperatures.
3) 365 x 120 daily cycles between day and night temperatures.

Extreme temperatures are given in BS5400 for 120 year return period. Annual and daily cycles can be estimated from TRRL 696, from which Figure 2 is copied showing histograms of daily temperature changes for a steel box girder. The combined histogram of Figure 2 approximates to a Rayleigh distribution with a r.m.s. value at about 13 degrees. Hambly et al (1982) show that a Rayleigh distribution of cyclic loading causes the same fatigue damage as uniform cycles of loading of about 80% of the significant value (average of highest 1/3) which is equivalent to about 110% of the r.m.s.: i.e. 14 degrees here. More detailed calculations can also be done for annual and daily ranges, but are unlikely to affect the conclusion.

The overall force on the viaduct cross-section is caused by the change in the average effective bridge temperature. However the stress range at any joint also includes the locked-in component from temperature differences. Using BS5400, TRRL 765 and TRRL 696, the viaduct in Figure 1 might be expected to experience the following cycles:-

| | Cycles | Temperature changes effective deg C | | difference deg C | | dT deg C |
|---|---|---|---|---|---|---|
| Extreme | 1 | 70 | + | 20 | = | 90 |
| Annual | 120 | 50 | + | 20 | = | 70 |
| Daily | 44,000 | 14 | + | 14 | = | 28 |

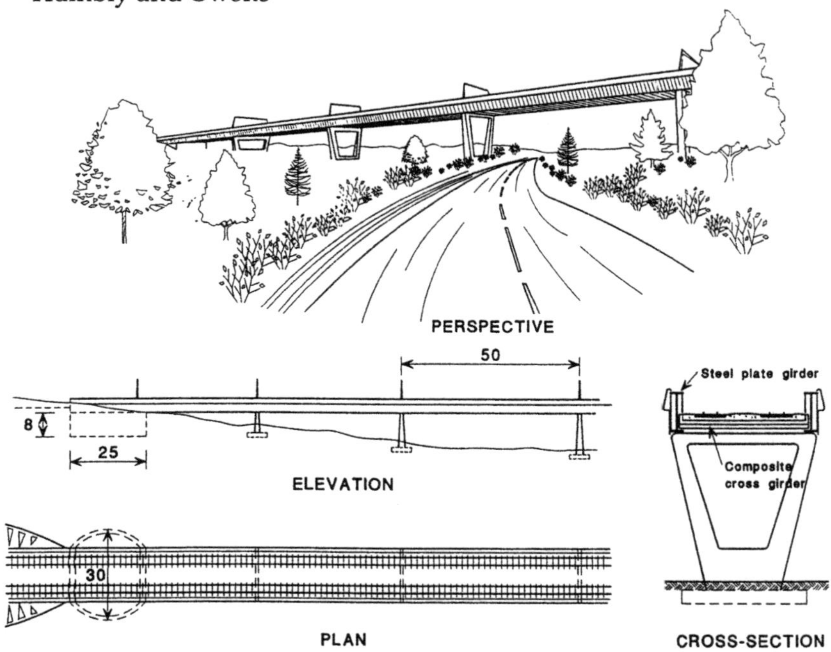


Figure 1 General arrangement for a jointless steel viaduct

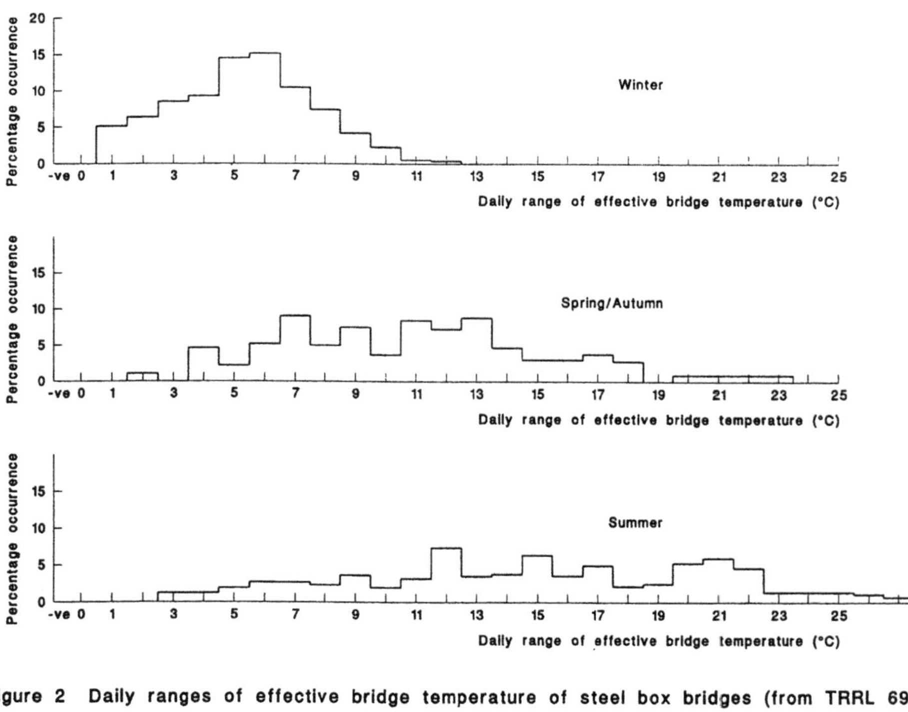


Figure 2 Daily ranges of effective bridge temperature of steel box bridges (from TRRL 696)

### 3.2 Thermal fatigue

Assuming coefficient of thermal expansion a=0.000012 per degree Centigrade and Young's modulus E=200,000N/mm$^2$, the temperature ranges of dT=28 deg C induce stress ranges ds of

$$\begin{aligned} ds &= dT\ a\ E \\ &= 28 \times 0.000012 \times 200{,}000 = 67\ \text{N/mm}^2 \end{aligned}$$

Assuming Class F welds, BS5400:Pt10 Figure 14 indicates that for a stress range of 67 N/mm$^2$ the permissible number of cycles is about N=2,000,000. Here the number of daily cycles is n=44,000: hence the fatigue damage from daily cycles would be:-

$$D = n/N = 44{,}000/2{,}000{,}000 = 0.02$$

Which is negligible as compared to the permissible damage of D=1.0. Similar calculations for the annual and extreme cycles of temperature also indicate negligible fatigue.

### 3.3 Live load fatigue

A preliminary estimate of live load stresses for acceptable fatigue design can be obtained from BS5400:Pt10 paragraph 9.2. Assuming RU loading the live load stress range could be up to

$$ds = k_1\ k_2\ k_3\ k_4\ k_5\ ds_o$$

as given in BS5400:Pt10. With Class F weld details the equation becomes:-

$$ds = 1.0 \times 1.0 \times 2.19 \times 0.89 \times 1.27 \times 40 = 100\ \text{N/mm}^2$$

### 3.4 Ultimate thermal force

It is assumed here that the viaduct has spans of 50m. RU loading on both tracks will cause a sagging moment in each girder at midspan of about 20MNm, when the span is loaded, and hogging moment of -10MNm, when adjacent spans are loaded. Thus the range of moment will be 30MNm. With stress range limited by fatigue to ds=100 N/mm$^2$, as above, the section modulus Z of each girder will need to be about

$$Z = M/ds = 30/100 = 0.30\text{m}^3$$

If the section is assumed to be 3.5m deep with radius of gyration r = 1.4m it will have moment of inertia I and area A of

$$I = 0.30 \times 3.5/2 = 0.52\text{m}^4$$

$$A = 0.52/1.4^2 = 0.27\text{m}^2$$

If the viaduct is fixed to the abutments when at the middle of its range of effective bridge temperature, the maximum rise of 35 deg C will cause a compression stress f and force P of

$$f = 35 \times 0.000012 \times 200{,}000 = 84 \text{ N/mm}^2$$

$$P = 84 \times 0.27 = 23\text{MN}$$

Thus the abutment must resist a longitudinal force from two girders of 46MN (4,600 tons), pushing or pulling.

The maximum compression stress of 84 $\text{N/mm}^2$ will contribute to buckling of the viaduct spans as a whole and to lateral buckling of compression flanges.

## 4 Abutment design

It is shown above that the abutments must resist longitudinal forces of 46MN in both directions. The abutment in Figure 1 is shown as an open box of diaphragm walls of about 25m x 30m with depth 8m. If the walls are 2m thick they will have weight 50MN and will enclose soil of weight 80MN. If the soil has angle of friction 35 deg the maximum sliding resistance would be

$$F_{sliding} = (50 + 80)\tan(35^\circ) = 100 \text{ MN}$$

If the soil of density 0.02MN/$m^3$ has coefficient of passive resistance of Kp = 6.0, then the a wall of width 30m and depth 8m could mobilise a passive force of

$$F_{passive} = 6.0 \times .020 \times (8^2/2) \times 30 = 115 \text{ MN}$$

The safety factor against sliding would be

$$SF = (100 + 115)/46 = 4.7$$

The maximum working coefficient of earth pressure would then be about $K = K_p/SF = 6/4.7 = 1.3$ so that the maximum movement, according to Hambly and Burland (1979), would be of the order of 0.001 of wall depth, i.e. about 8mm (once in 120 years). More precise calculations could be undertaken using finite elements.

## 5 Pier design

It is anticipated that the completed viaduct will be fixed to the piers in order to dampen noise. It is possible that the viaduct would be supported on sliding/rolling bearings during construction and then fixed to the piers after the viaduct has been attached to the abutments.

Under the action of longitudinal forces from braking and traction the jointless viaduct will provide a stiff line of reaction to the abutments, while the piers will be relatively flexible. Consequently there is no point in making the piers stiff, and their flexibility will be advantageous in accommodating the effects of any variations in temperature along a structure. The total design longitudinal loads are small as compared to the temperature load of 46MN calculated above.

## 6 References

Burke, M. P. (1990) "Integrated bridge design is on the rise." **Modern Steel Construction AISC**, July 1990.

Hambly, E. C. (1978). "Temperature distributions and stresses in concrete bridges"; **The Structural Engineer**, 56A, May pp 143-148.

Hambly, E. C. and Burland, J. B. (1979). "Bridge Foundations and Substructures"; Her Majesty's Stationery Office.

Hambly, E. C., Edwards, A. J., Kohli, C. & Miller, B. L. (1981). "Fatigue considerations for ocean towage". OTC 4163, 13th Offshore Technology Conference, Houston.

Hambly, E. C. & Nicholson, B. A. (1990). "Prestressed beam integral bridges", **The Structural Engineer**, Vol 68, No 23, December.

Hambly, E. C. (1991). "Bridge Deck Behaviour", 2nd Edition, E & FN Spon, London.

Hambly, E. C. (1992). "Integral bridge abutment details in practice and theory", Seminar on Bridge Design for Durability, Transport Research Laboratory, Crowthorne.

TRRL 696 (1976), "Bridge temperatures estimated from the shade temperature", by M. Emerson, Transport and Road Research Laboratory, Crowthorne.

TRRL 765 (1977), "Temperature differences in bridges: basis of design requirements", by M Emerson, Transport and Road Research Laboratory, Crowthorne.

# REPORT ON SESSION FOUR

S. SHANMUGAM

The first three papers of this session discussed the Danish, French and Italian experiences with continuous and integral bridges.

Ejgil Veje briefly described some of the recent major Danish bridge projects, which are mainly of long span continuous construction. For a suspension bridge he said that they have designed a 2700 m long continuous girder between the two anchor blocks, the longest jointless length ever achieved in bridge design. He added that such large spans have necessitated the development of bearings and expansion joints to accommodate the increased movements involved.

Jean Muller showed interesting slides describing the prestressed concrete continuous bridges built in France and how they coped with thermal effects.

Professor Cabrera described various types of continuous viaducts built in Italy in both precast and cast in place, explaining the design and construction processes involved. In response to a query about the longitudinal box girder being constructed in two halves, the Author said that each half was cast separately and the units were connected in the transverse and longitudinal directions after positioning, and that there were no problems.

In introducing his paper, Mr Paterson referred to several continuous arch bridges built between 1850 and 1900; continuous span bridges constructed in the 1920s and prestressed concrete beam bridges built in the 1970s. He reported that a small number of integral bridges were built in the 1960s and 1970s, all of which have performed very well.

He briefly discussed the contents of the proposed Department of Transport Advice Note on Design for Durability which is intended to give guidance on good practice. It will be accompanied by a Departmental Standard which will wherever possible require all bridges to be continuous over intermediate piers and fully integrated with abutments for bridge lengths up to 70 m. It will also include buried structures, and arch bridges both in concrete and masonry, where they stand to benefit from reduced maintenance and enhanced durability. He concluded that the re-establishment of arch bridges has now brought the whole development to full circle.

Dr Owens introduced the paper on Jointless Steel Viaducts and demonstrated that for lengths up to 200 m steel viaducts for high speed railways can perform satisfactorily without expansion joints at either abutment. For such steel viaducts to be compatible with welded railway track during temperature change, he pointed out that massive anchored abutments would be required. He suggested that economy might be achieved if the structure was shielded against extreme temperatures.

# PART FIVE

# LONG-LENGTH CONTINUITY

# 15 THE DESIGN AND PERFORMANCE OF CONTINUOUSLY REINFORCED CONCRETE PAVEMENTS

F. FUCHS
Belgian Road Research Centre, Brussels, Belgium
R. DUMONT
Ministry of Infrastructure and Transport, Wallonia, Belgium

**Abstract**
The paper describes the use and behaviour of continuous reinforced concrete pavements in Belgium, in particular the variants that have been used in constructing CRC pavements at overpasses without interruption.
Keywords : Bridge, Deck surfacing, Continuous, Reinforced concrete.

## 1 Continuously reinforced concrete pavements in Belgium

The continuously reinforced concrete (CRC) technique has been used on a large scale in Belgium, particularly for motorway construction since the early seventies (ref. 1). Over 18 million $m^2$ of CRC pavements were constructed since 1970.

The main advantage of this type of pavement is the absence of transverse joints, which makes it possible to construct lasting pavements requiring virtually no maintenance. Cracking is controlled by continuous reinforcement located midway between the top and bottom of the slab and designed in such a way that a fine network of cracks (average crack spacing between 1 and 3 m) is created which does not imperil the good performance of the pavement. The main factors influencing the crack pattern are :

- the percentage of longitudinal reinforcement related to the yield point of the steel and to the tensile strength of the concrete,
- bond strength between the concrete and the steel.

In Belgium we use deformed bars with a yield point of 500 MPa. The percentage of longitudinal steel is 0.67 % with respect to the cross section of the 20 cm thick concrete pavement (16 mm diameter bars with a spacing of 15 cm). Initially, from 1970 to 1976, the percentage of longitudinal steel was 0.85 % (18 mm diameter bars every 15cm), but after a survey of the first motorways constructed it was reduced to 0.67 % mainly in order to increase average

*Continuous and Integral Bridges.* Edited by B.P. Pritchard. Published in 1994
by Taylor & Francis, 2 Park Square, Milton Park, Abingdon, Oxon, OX14 4RN. ISBN: 0 419 19030 9.

crack spacing and obtain a more homogeneous crack pattern.

The reinforcement consists either of all-welded fabric fully assembled in the factory, or is composed of transverse bars provided with supports welded to them in the factory, the longitudinal bars being subsequently fixed to these transverse bars on the construction site by clipping the ends of the supports. The transverse bars are placed at an angle of 60° with the longitudinal bars in order to counteract the tendency for cracks to develop at transverse bars perpendicularly to the centre line of the road.

The performance of most CRC pavements constructed since 1970 can be rated as excellent, especially when considering the little maintenance required. A survey recently conducted by the Ministry of Public Works and the Research Centre of the Cement Industry has shown that there is practically no corrosion of the longitudinal reinforcement, the loss in cross section of the steel being generally less than 1 % after ten to twenty years of service (ref. 2).

It has also turned out that this type of pavement is able to resist subsidence in the embankments of bridge approaches, the structure and evenness of the pavement being restored by injecting a cement grout.

The only defects that have occurred systematically are :

- construction joint failure during hot summers, with some cases of blow-up of the pavement due to the poor quality of the concrete in the vicinity of the joint,
- punch-outs at the outer edge of the pavement on some recently constructed motorways, due to inadequate drainage at the interface between the pavement and the road base and the poor quality of the sub-base material.

## 2 Continuously reinforced concrete roads at overpasses

Interrupting CRC pavements at overpasses raises problems :

- **of construction** : because of the close spacing of engineering structures on Belgian motorways, concreting has to be stopped at frequent intervals and transition zones have to be made in bituminous materials, which is not a very rational working method;
- **of cost** : constructing anchor abutments and transition zones at each engineering structure is expensive;
- **of management** : whereas CRC pavements need little maintenance for more than twenty years, joints and bituminous surfacings require interventions every five to seven years.

There are three variants in constructing CRC pavements at overpasses.

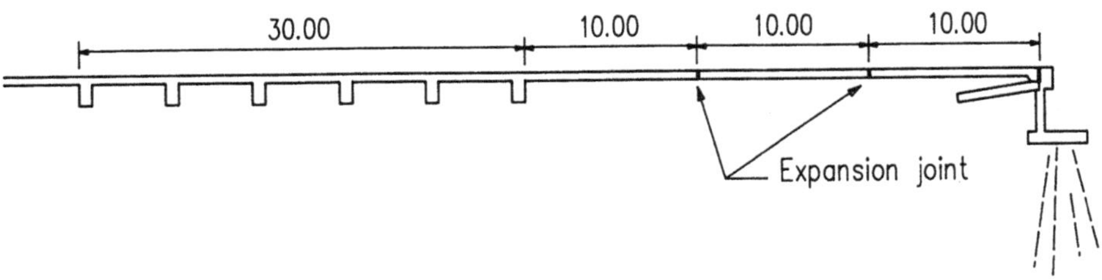


Fig. 1 Interruption of a CRC at an overpass approach.

a) CRC on overpass approaches
At structures not designed to resist forces applied by CRC, the pavement is interrupted by an anchor abutment. This abutment is located at a distance such that the foundation of the structure is shielded from thrusts produced by expansions of the concrete pavement.

The transition zone between the anchor abutment and the structure is constructed in jointed reinforced concrete slabs.

This procedure makes it possible to reduce stresses and strains in the bridge joint, but retains all the disadvantages involved in interrupted concreting.

b) CRC on conventional structures
**Frame bridges**
The stiffness of this type of structure enables the CRC pavement to be continued on the bridge as far as the design engineer has allowed for a permanent load by that pavement (500 kg/m² for 20 cm of CRC) and a horizontal force equalling half that load (250 kg/m² for 20 cm of CRC).

The bridge has to be constructed with an extra ± 10 % of concrete to resist those forces, but this is partly compensated by the possibility to omit the floating slabs.

**Girder bridges**
With span lengths greater than ± 20 m frame bridges are no longer cost-effective, and CRC pavements are constructed on girder bridges.

A significant horizontal force is applied by the pavement to this type of structure, which is less rigid than a frame bridge. This horizontal force is reduced by incorporating between the CRC and the waterproofing system of the structure :

- either a double polyethylene sheet reducing the horizontal force to ± 20 % of the vertical force applied by the CRC,
- or sliding sheets reducing the horizontal force to ± 10 % of the vertical force applied by the CRC.

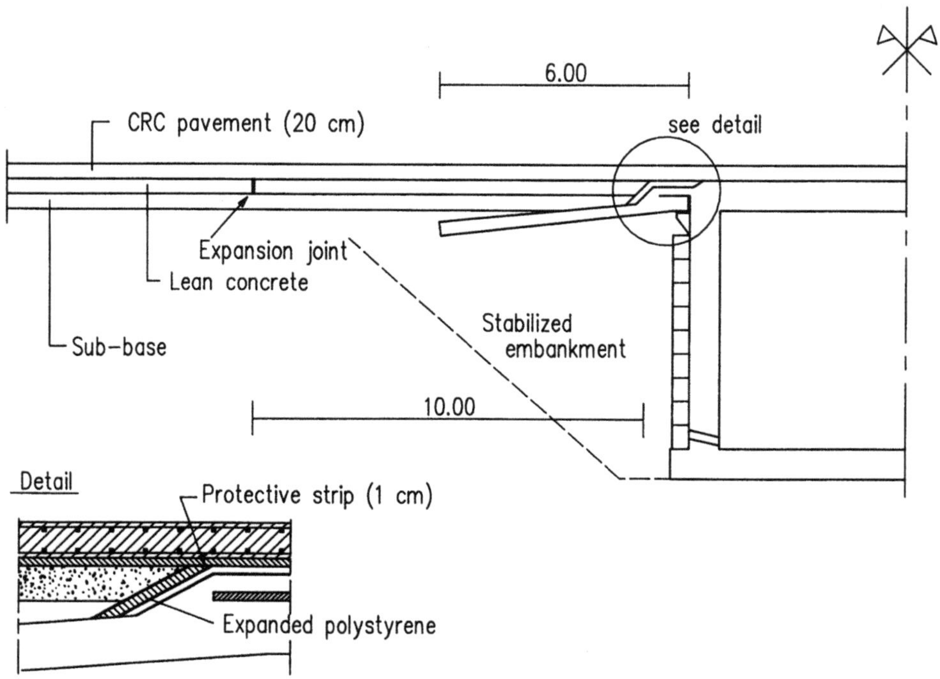


Fig. 2 CRC on frame bridges.

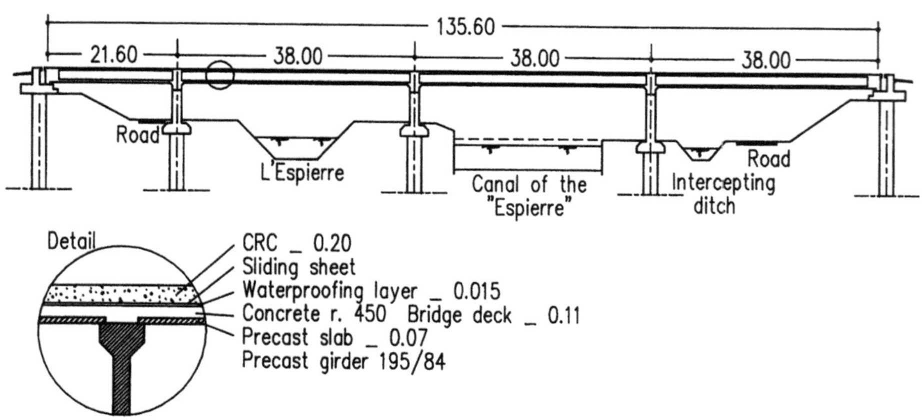


Fig. 3 CRC on girder bridges.

**c) CRC on/in bowstring bridges**
The load superimposed by a concrete pavement on the bridge can be reduced if the CRC pavement is made an integral part of the bridge structure.

Neoprene bearings between the transverse girders and the precast slabs underlying the CRC pavement allow easier movement of the bridge with respect to that pavement. The pavement slab is laid with a slip-form paver.

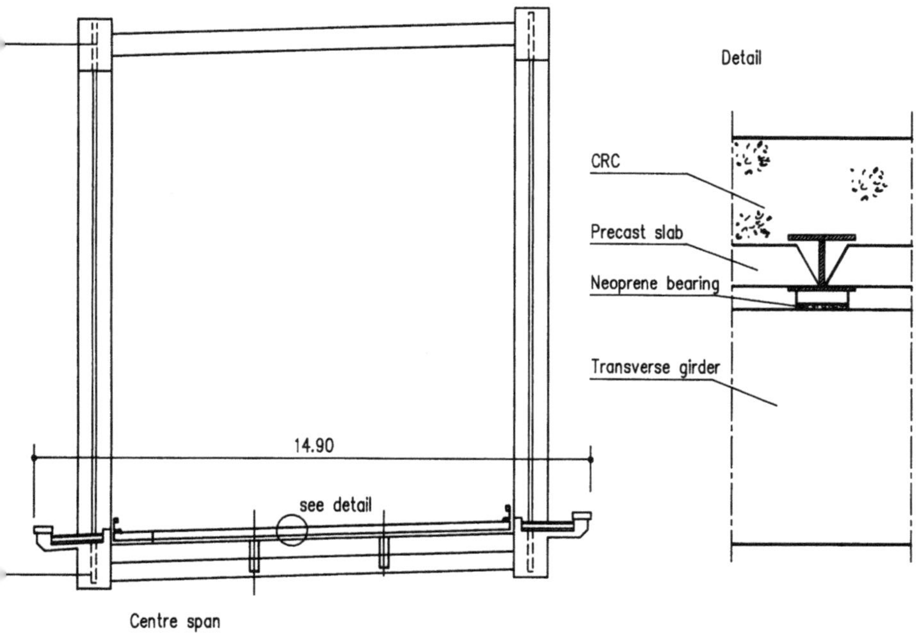


Fig. 4 CRC in bowstring bridges.

## 3 References

Y Dechamps, R De Paepe, P Dutron - Belgian experience with continuously reinforced concrete pavements. International Conference on Concrete Pavement Design, Purdue University, 319-334, 1977.

K Verhoeven - Cracking and corrosion in continuously concrete pavements. 5th International Conference on Concrete Pavement Design and Rehabilitation, Purdue University, 201-209, 1993.

# 16 EXPERIENCE IN THE USE OF CONTINUOUS WELDED RAILS ON THE NETHERLANDS RAILWAY NETWORK

G. BIER
Ingenieursbureau, Nederlandse Spoorwegen,
Utrecht, The Netherlands

Abstract
Modern tracks consist of continuous welded rails. Temperature changes introduce forces in the rail. At the ends of the CWR there are anchor lengths, in which these forces are reduced to nil by a number of sleepers. Because of the elasto-plastic resistance behaviour of the sleepers in the ballast, big displacements of the rail ends occur, for which special expansion devices are needed.

Bridges under the track give extra stresses in the rail, caused by both temperature elongation and shortening of the bridge deck, and by traction or braking forces from trains running over the bridge. These effects ask for reduced bridge deck lengths, and minimum stiffness of bridge bearings and foundation.
Keywords: continuous welded rail, anchor length, expansion joints, bridge deck lengths, bridge bearing's stiffness.

## 1 What is CWR on embankment

One of the big changes in Permanent Way engineering has been the introduction on a large scale of continuous welded rail (CWR). Today most of Netherlands Railway's network is fitted with this low maintenance type of track.

When you get rid of the fishplated rail joints that were there every 18, 24 or 30 meters, the question is how the mechanics of the system work.

For any piece of rail somewhere in the middle of the CW-track we can see that it is in equilibrium between the two surrounding ones. It follows that in this middle section of the rail stresses are being introduced. No extra longitudinal force is needed from below i.e. from sleeper and ballastbed.

This situation changes when we approach the end of the CWR section. There must be an end at CWR at places where the internal rail forces, caused by the CW, cannot be withstood. This can be the case at complex swith & crossing spots, or before narrow curves, mine settlement areas and, particularly in the Netherlands, at movable bridges.

*Continuous and Integral Bridges.* Edited by B.P. Pritchard. Published in 1994
by Taylor & Francis, 2 Park Square, Milton Park, Abingdon, Oxon, OX14 4RN. ISBN: 0 419 19030 9.

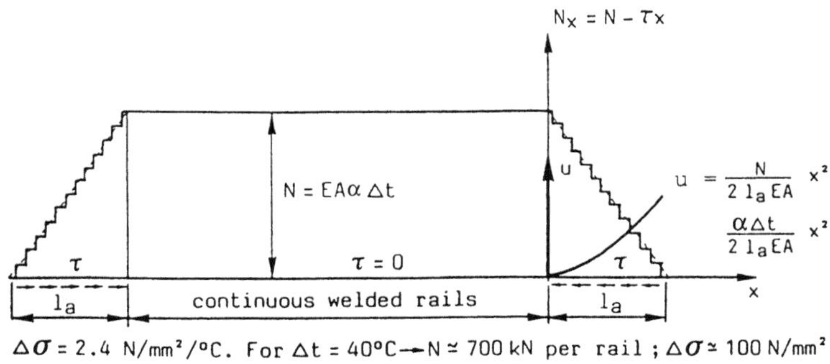


Fig.1. Distribution of temperature force in CWR

At the end of the CW rail we know for sure that there is no longitudinal force. A piece of rail at little distance from the rail end will be pushed away by a bigger force coming from the CW-side than from the side of the expansion joint. These pieces in the so called anchor- (or breathing-) length can only be brought into equilibrium by an extra force delivered by the sleeper. So here, the sleeper does have to resist certain longitudinal forces from the CWR.

Since the longitudinal forces in the CWR can reach values of 500 to 700 kNs per rail (stresses up to 100-N/mm²), it is impossible to stop these forces by one sleeper or fastening. A number of sleepers together have to resist the CWR forces. By doing so, every sleeper is pushed through the ballast before it comes to its maximum resistance value.

This means that at the end of the CWR the rail can undergo quite some variation in length, which reduces the internal rail force to zero at the end. But this is accompanied by such displacements at the end, that special rail expansion devices (adjustment switches) are to be used.

The movement in the ballast occurs when rail forces reach a certain level. The bigger the temperature changes are, the more resistance is needed. Frequent movements of the sleepers through the ballast are unfavourable for the packing of the ballast and thus influence both ballast resistance (longitudinal and lateral) and the track's level and alignment. Which reduces riding quality and increases the need for track maintenance work.

One other mechanics aspect is the risk of track buckling at higher temperatures.

Considering a straight bar that we put under longitudinal pressure, we know that at a certain level of internal stress, the bar buckles. Holding this bar in the middle, raises the critical longitudinal buckling force to the square. Holding the bar on two more points, again increases the critical buckling force resistance. The miracle of CWR is that normal sleepered track can, under specific conditions, resist the forces that occur in our climatical circumstances.

Rail temperatures can, in the Netherlands, as probably is the case in the United Kingdom, vary between minus 25 and plus 55 degrees Celcius. We always fasten the rails at

the sleepers at the same, so-called neutral (stress free) temperature. Since track buckling is more critical then high tension stresses in the rail are, this temperature is chosen above the mean value of the temperature limits. At NS it is 25°C.

## 2 What role does the bridge play?

When we look back on the situation in the middle of the CWR, we see the influence of the bridge coming to the track in the form of an extra longitudinal displacement of the base of the ballastbed. This leads to forces coming from ballastbed to sleeper and finaly rail, resulting in a disturbance of the continuous CWR stress level. At the movable end of the bridge, rail stresses increase, whereas over the fixed bearing area they get reduced.

Three limits are to be considered (in order of importance):

1. the track buckling resistance.
2. the rail (tension) strength.
3. displacements in the permanent way construction, especially of importance when they occur often.

Since the cross section of the bridge deck is so much bigger than that of the rail, no specific attention is paid to stresses in the deck. The more however, we must consider the forces in, and displacements of the bridge deck supports, in particular when rubber bearing blocks are used.

And, not to forget, do the same for the rail supports.

## 3 The loads

For the interaction between track and bridge mainly two characteristic longitudinal loads, must be considered:

1. Temperature effects
   Caused by the climatical influences the bridge decks expand or shorten. One can distinguish between the variation over the year (slow) and over the day

(quicker, but still of static nature). These effects occur mainly in the unloaded track situation.

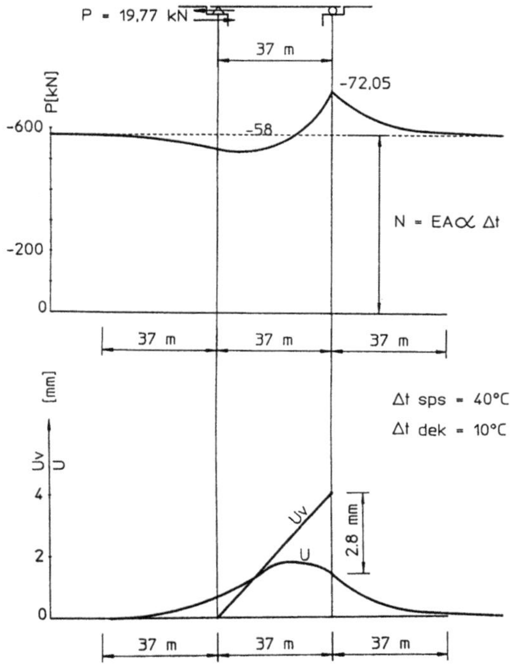


Fig.2. Temperature effects

2. Traction (maximum 33,3 kN/m') and breaking forces (maximum 20 kN/m').

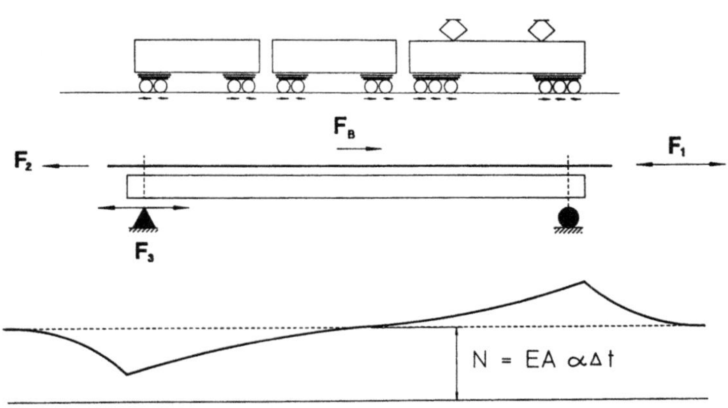


Fig.3. Traction and braking effects

These forces are quasi-static to even dynamic of character. Especially long freight trains can give big forces. German surveys have shown that the biggest force occurs at the last 0,2 seconds period of the braking manoeuvre of a train. Every car will stop at the same time which leads, at long trains, to a very big peak load of dynamic kind. In flexible bridges this can cause very big forces in the continuing rail and displacements of bridge to rail and rail to embankment.

## 4 Between track and bridge

Between rail and bridge there is always a more or less elastic element. At ballasted tracks, favoured by permanent way engineers, it is mainly the ballastbed that acts like an elasto-plastic medium. The resistance values of a ballastbed are depending on the shape of the stones, the ballast depth below the sleeper, the type of sleeper, the fastening properties, and, very important, the compaction of the ballast. The compaction is much lower after maintenance operations or when the ballast moves frequently a lot.

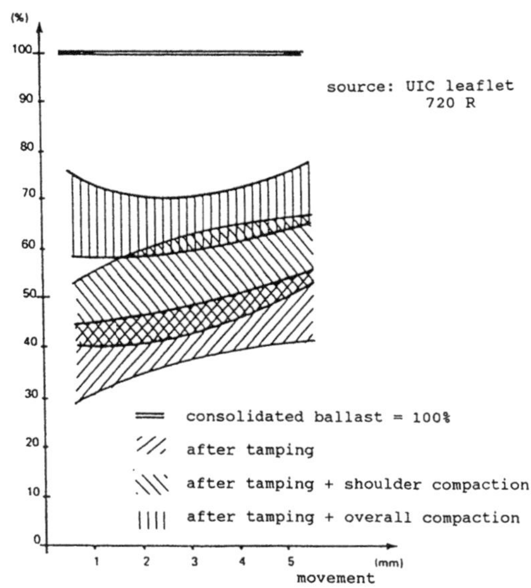


Fig.4. loss of lateral ballast resistance due to maintenance operations

When any type of direct rail fastening is to be used, investigation must be done to the elasto-plastic behaviour of it. The advantage of direct rail fastening can be the more constant force and displacement properties. Less favourable is the greater stiffness and reduced displacement capacity. This results in general in less possibilities for making CWR over bridges.

## 5 Rail expansion joints

In order to compensate bigger displacement between two following rail sections, special devices have been developed. They are characterised by a joint that crosses the rail as longitudinal as possible. Thus, two tongues are shaped.

Like in turnouts, the passing wheel first runs at one tongue, and then gradually starts to run at the other one. This taking over goes more smoothly, the flatter the angle of the joints to the track direction is.

Although these constructions are much more expensive, they often cannot compete with the costs of bridge deck joints.

Another point, however is, that these devices need much more inspection and maintenance. Also they have, in general, shorter life times, which means that more often than the surrounding rails, they have to be replaced.

Expansion joints interrupt the CWR, and this, as we have mentioned before, creates anchor length areas, with all their disadvantages such as movement, wear, disturbance of track level and alignement, and so on.

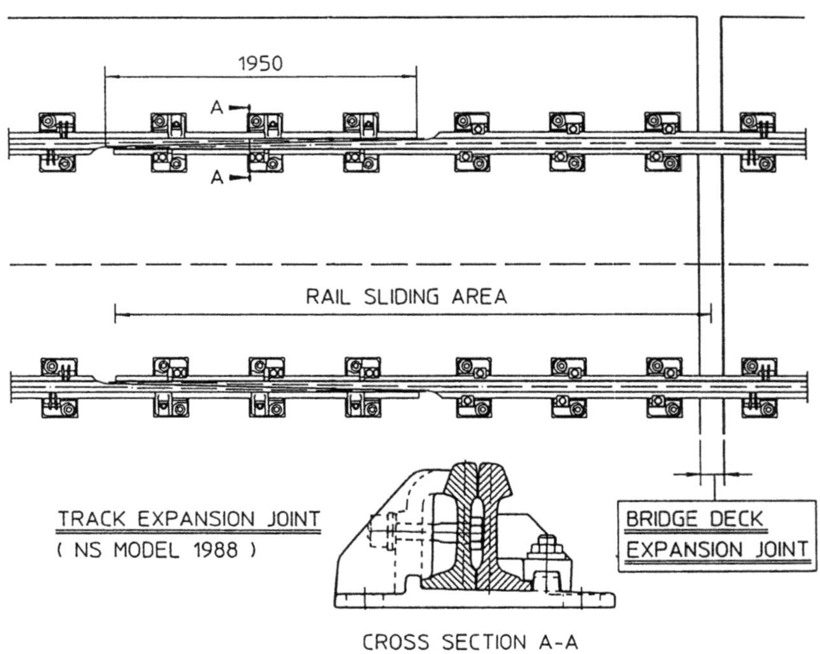

## 6 Prolis computer programm

Since, in the Netherlands we have only little differences in level, many bridges are designed with direct railfastening because of its low construction depth. Because those are more influencing the interaction between bridge and track, a lot of calculations have been done. The mathematical possibilities have grown much, and so we have finally seen the development of a new computerprogramm, called Prolis. It was developed by two students of the Technical University of Delft, accompanied by several people from Netherlands Railways. Based on the finite elements principle many properties in the design can be varied to calculate several different variants.

## 7 UIC 7J18

By UIC, the International Union of Railway Administrations, a working group (called 7J18) was established, with the aim to bring forth a design leaflet for railway bridges. Both people from the Bridge Departments (BR,DB,SNCF) and Permanent Way Departments (NS and SNCB) took part in this group. After several years of exchange of thoughts last Januari a general agreed proposal was offered to the charging Bridge commitee (UIC 7J). Discussion in the Permanent Way committee (UIC 7G), is still under way, where especially the matter of maximum rail forces and displacements is still in view. Estimated limits for additional railstresses caused by bridges are 72 $N/mm^2$ for pressure and 92 $N/mm^2$ for tension. Movement of bridge decks under CWR should stay under 5 mm.

## 8 Experiences at the NS network

From the earlier days of the railway we had many big river crossing bridges, all fitted with ballastless track, and with use of expansion joints.

After the second world war, and especially from the mid sixties, many new viaducts and bridges have been built. In our more and more dense populated country, the designs became more and more complicated. Not only straight, but also curved with pile patterns sometimes very complicated. Trough bridges skeweing at very sharp crossing angles to the infrastructure beneath, enable free crossing of traffic flows, but ask severe design work.

Until the mid eighties many bridges with direct railfastening have been designed and built. With those, after ten to twenty five years of experience, we have gained knowledge about what is possible.

Not always the experiences were free choice. For example, just in the old city of Delft, we have an 800m long

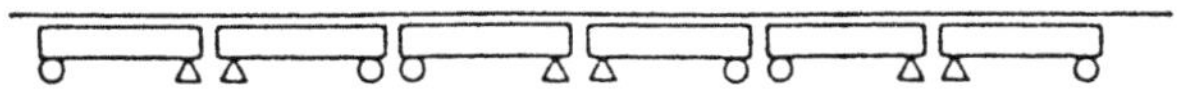

Support arrangement

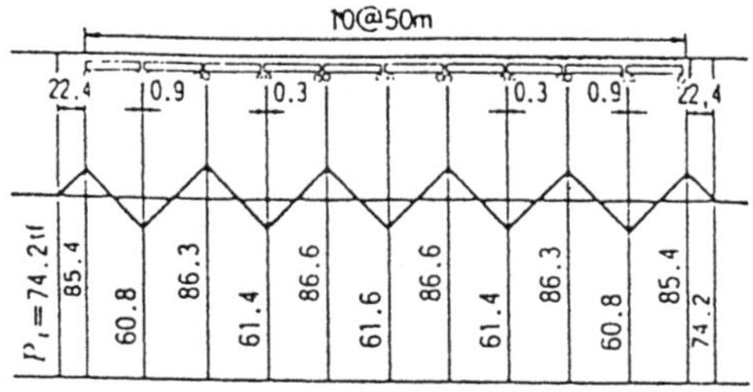


Fig.5. Longitudinal rail forces at long bridges

viaduct, double tracked. About half the number of trains is passing at a speed of 100 to 110kms/hour. The other trains pass at low speed and are braking for a stop in Delft Central Station. At the end, closest to this station, there is a curve of about 550m radius. The viaduct exists of 17 sections of 55m length each, and the rails are directly fastened to the decks. When the viaduct was put in use, there was a rail expansion joint at every bridge joint. Those however, already after short time gave big troubles with maintenance, especially in the narrow curve. A solution was found, without computer facilities, but still an analytical model showed that it was possible to make CWR over the viaduct. It was necessary to create anchor length over both end abuttments, in order to prevent the first deck being pushed from its piles.

All calculations that have been done for the so called Delft's Regular Railfastening show in fact that a deck length of 55 m is to long. However, based on many years of experience, we have practically considered the 55 m to be possible. Probably there is a favourable effect of some relaxation in the railfastening.

## 9 Today's practice

Our more recent practice however is that we try to get ballasted tracks designed on the bridges enableing the bridge designer to cross over dilatating lengths of 90 m for concrete and 60 m for steel bridges. These deck lengths are advantageous, in particular in the complicated situations earlier mentioned.

For the PW engineer especially the smoothening of the transition to the adjacant embankment, and the possibility of continuing maintenance operations over the bridge are of great advantage.

When however, bridge decks have to be so long that expansion joints are needed anyhow, the fixed position of those joints in direct railfastening is preferable.

# 17 CONTINUOUS ARCHES

D.R. THOMPSON and G.W. CLARK
British Rail Research, Derby, UK

Abstract
Masonry arch bridges are constructed entirely without the use of movement joints. This paper looks at the nature of masonry arch bridges and highlights their main attributes. The main strengths and weakness of this type of structure are considered in the context of the continuous bridge deck.
Keywords: Masonry, Arch, Viaduct.

## 1 Introduction

Masonry arches are one of the oldest forms of bridge construction. The arch was used by many of the ancient civilisations but the Romans are notable as many fine examples of their arches have survived and some still carry traffic 2000 years after they were built [1].

Masonry arches of all types were extensively used in the construction of railways, particularly in Britain, and many of these structures are still in use today. Masonry arches are used in structures ranging from simple culverts to mile long tunnels and viaducts.

Masonry arches may be constructed from many different materials such as brick, stone or mass concrete.

An arch is a pure compression structure; it carries loads by means of individual blocks (voussoirs) pressing against each other. This form of construction is fundamentally stable, and has proved very durable.

This paper will look briefly at the way in which the arch works. It will show how individual arches may be strung together to form long bridges without the need for construction joints. Finally, a comparison of the arch and beam for continuous bridges will be made.

## 2 How an arch works

Figure 1 shows an idealized arch with the main elements

*Continuous and Integral Bridges.* Edited by B.P. Pritchard. Published in 1994
by Taylor & Francis, 2 Park Square, Milton Park, Abingdon, Oxon, OX14 4RN. ISBN: 0 419 19030 9.

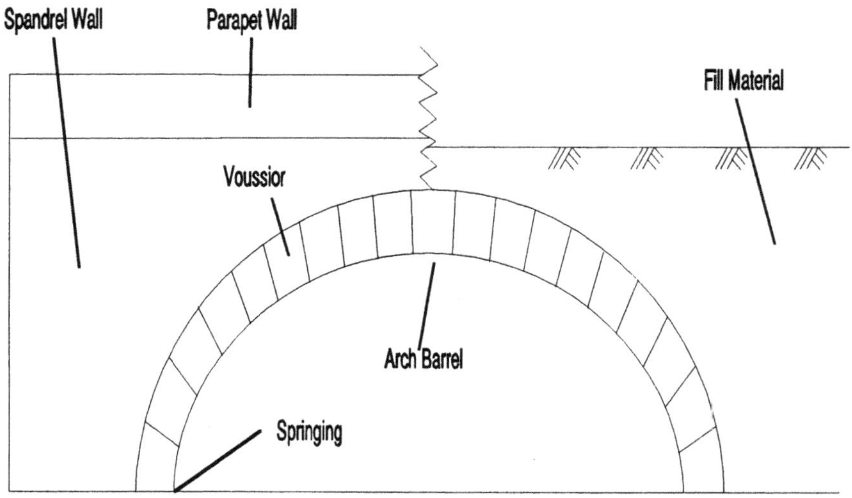


**Figure 1:** Idealised arch

detailed. Arches may be constructed from simple single elements such as shaped stones or bricks (masonry) or from mass concrete. The vast majority of the arches on British Rail (BR) are constructed from masonry. Masonry (and mass concrete) is characterised by its inability to sustain tensile forces.

The arches considered here are, by definition, pure compression structures with no ability to resist internal tensile forces. The use of reinforced concrete or steel in an arch would lead to a structure with very different characteristics and is outside the scope of this paper.

The main load bearing element of an arch is the barrel. The barrel can be considered as a series of elements (Figure 2). Each element supports its own dead weight and the dead weight of the fill above it. It is held up by components of the forces exerted on it by the elements on either side of it. Consequently the dead load stresses from the self weight of the fill and the barrel material are carried by the barrel. The centre of action of the stresses in the barrel is called the thrust line.

These dead load stresses are crucial to the stability of the arch. They act in a similar manner to the pre-stress in a pre-stressed concrete beam in that they enable a material with no tensile capability to carry loads which would otherwise produce tensile stresses. The

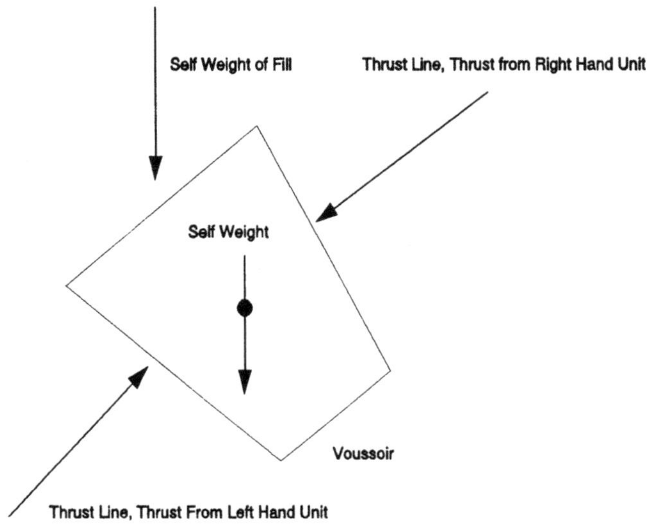


**Figure 2:** Forces on a barrel element

greater this 'dead load prestress' the greater the live load capacity, unless the limiting compressive stress of the masonry is exceeded. The position of the thrust line need not be coincident with the centroid of the barrel, although, since the thrust line must remain within the barrel thickness at all times, the closer it is to the centroid, the more load the arch can carry for a given barrel cross section and material. The geometry of the arch is important here, as some shapes (e.g. elliptical) lead to the development of a thrust line that is close to the surface of the barrel even under dead load only.

If an external force is applied to an arch, then the thrust line alters its position in the barrel, to maintain the static equilibrium of the structure. As the thrust line moves away from the neutral axis of the barrel, bending moments are introduced. The further the thrust line moves then the bigger the bending moments will become. As the thrust line moves outside the middle third of the barrel thickness tensile stresses try to develop at the opposite surface. This leads to tensile cracking of the masonry and a redistribution of the compressive stress across the remaining part of the section. As the cracked zone grows, the area of material available to resist the thrust in the barrel decreases and the compressive stress increases. This increase in local stress may cause crushing. Once all of the material across the section has either cracked or crushed no further increase in the loading at that section can be sustained and a plastic hinge forms. Due to its shape, an arch can still function with three such hinges in most cases. When a fourth hinge forms, however, the arch becomes a mechanism and collapse occurs.

The precise points at which the hinges develop depends on the geometry of the arch and the position of the loading. It is the development of these hinges which controls the ultimate capacity of the arch in most cases.

The behaviour of a real arch is complicated by the interaction between the barrel and the surrounding fill. As the arch barrel deforms, the whole structure (both barrel and fill) distorts as a homogeneous whole. Movements in the fill cause stresses within the fill itself and between the fill and the external surface (extrados) of the arch barrel. This interaction between the soil and the structure makes the analysis of arches a complex problem. The behaviour of arch fill material is not fully understood. In practice it is often difficult to determine the nature of the fill and its physical properties.

The durability of the arch comes partly from its reliance solely on compression for its strength. This means that the arch is inherently stable, even if individual elements deteriorate or crack. The compressive stresses within the arch barrel will tend to close up many cracks. Other faults can be accommodated by means of a re-alignment of the thrust line and changes in the shape of the arch.

## 3 How a viaduct works

Figure 3 shows the form of a typical viaduct. A masonry arch viaduct is a series of simple arches on piers. This form of construction allows a bridge of unlimited length. Each span can be different from those abutting it both in span and shape.

The way in which the viaduct behaves under load depends on the slenderness of the piers. Stocky piers can carry both vertical and horizontal loads directly to the foundations. This means that deflections from one span are localized and are not carried over to the adjacent spans. In many cases the piers are quite slender and cannot carry the entire horizontal load from one arch span. In such cases the load is also carried by adjacent spans as the loaded arch barrel deflects and changes shape. This pier sway away from the load causes upward deflections in the adjacent spans, which are resisted primarily by the dead weight on those spans but, also by the pier stiffness. Consequently the live load

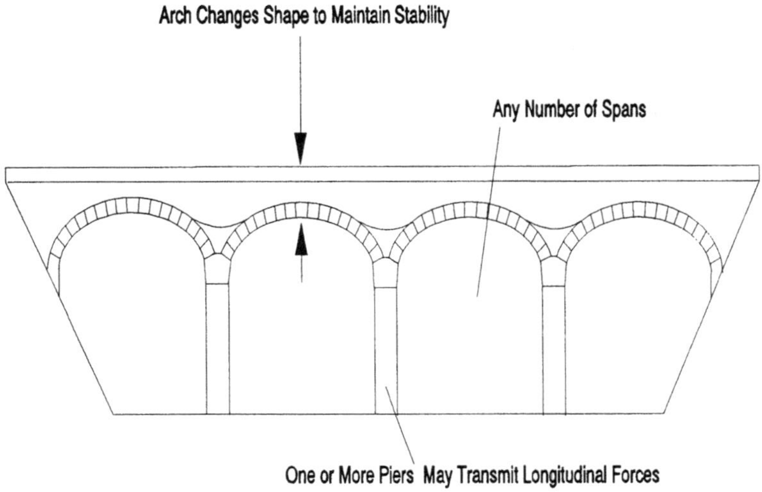


**Figure 3:** A typical viaduct

deflection of one arch span causes reduced movements in the adjacent arch spans. This effect is repeated in the next arch and so on until the movement, and therefore the stresses, decay away completely.

The inability of slender piers to carry all of the out-of-balance horizontal forces means that such viaducts are difficult to construct one span at a time and are susceptible to progressive collapse if a span is removed. For this reason stocky piers are usually built into such viaducts every five spans or so.

Masonry (and mass concrete) viaducts have no need of movement joins. Both loading and thermal movements, are absorbed by the structure in the form of changes in the shape of the individual arch barrels.

The analysis of masonry arch viaducts is a very complex operation. Problems caused by the fill/barrel interaction and the behaviour of the fill, for a single arch, are compounded by the ability of the supports of each individual arch to move independently. The failure mode of masonry viaducts is considerabley more complex than single arch spans; the collapse mechanism frequently involving hinges in several spans and piers. The effect of the spandrels is also more pronounced in multispan arches than in single span arches [2].

British Rail Research have developed the Masonry Arch Finite Element Analysis (MAFEA) [3] program to cope with the complex problems presented by arch analysis. MAFEA has been in use for single spans for some time. This program will, in time, be extended to cover multi-span structures.

As with single span arches, viaducts act as a homogeneous entity. Consequently longitudinal forces, such as braking and traction, are distributed over the whole length of the structure and dissipated. In the case of viaducts with stocky piers these forces can be expected to pass directly into the supports.

Figs 4 and 5 show the Culloden and the Harringworth viaducts. The Culloden viaduct is good example of the continuous brick masonry viaducts that can be seen all over the UK. Note the enlarged span to carry the structure over a road. The Harringworth viaduct shows the considerable length of structure that can be achieved.

## 4 Comparison with continuous beam structures

Bridges based on the simply supported beam principle are probably the most common form of modern construction. This type of structure is very simple to design and to analyse. Joints are used to divide the bridge into simple elements for analysis. This compels deflections to occur at predictable places throughout the bridge.

Recent experience, particularly with highway bridges, has shown that these joints can be difficult to maintain and inspect. The decay of the material around joints often leads to expensive repairs and even the need to replace the structure completely.

Arches do not need movement joints. Consequently they avoid this weakness, which is inherent in the design of some modern bridges.

In modern bridge decks, expansion joints are often used to remove the stresses that would be induced if temperature-related expansions and contractions were restrained. However, longitudinal braking and traction forces, must then be carried across the these joints.

Anchoring of the bridge deck against longitudinal forces is normally achieved by means of a small number of fixed bearings. This means that the braking and traction loads are concentrated at a few points, leading to high local stresses. In a masonry (or mass concrete) viaduct the resistance to longitudinal forces is distributed throughout the length of the structure and there is no concentration of force at joints.

**Fig 4:** Culloden Viaduct

**Fig 5:** Harringworth

## 5 Discussion

Large masonry arch bridges are no longer being constructed for a number of reasons. They tend to be very heavy and consequently they may require large foundations. Arches may be expensive to construct and require labour with skills that are not readily available in today's market. Masonry arches are not amenable to rapid construction and must be built in-situ due to the need for centering. The former disadvantage could be largely overcome by use of mass concrete instead of masonry but the latter applies equally to both materials. The curved arch profile reduces headroom adjacent to the springing. Arches are not suitable for large single spans unless a large construction depth is available and consequently long arch bridges have to be constructed as multi-span viaducts. Thus the designer must accept a large number of intermediate supports.

The difficulty in analyzing arch structures and in predicting their performance can also deter some engineers. Designers have become used to the more rigorous solutions available for more modern bridge forms.

However there are also considerable advantages with the arch form of construction which should not be overlooked. They have proved their durability on railways over the last 150 years and have been in use for canals and roads for considerably longer. They have proved to be low maintenance structures, despite the fact that many are now carrying considerably higher loads then they were originally designed for. They have a good load carrying capability and they can be built to form viaducts of any length without the need for movement joints.

Compared to modern construction forms, masonry arches are simple in concept. This combined with their low maintenance needs and proven 120 year plus life expectancy should lead to favourable whole life costing.

The advantages listed above are achieved by virtue of the combination of the simplicity of form and materials. Masonry arch structures do not require reinforcing steel or complex components such as bearings. This simplicity of form and materials means that they are not likely to be prone to unexpected failure mechanisms as is the case with reinforced and prestressed concrete. In addition, maintenance is simple and no special components, materials or techniques are needed.

## 6 Conclusions

The following points may be drawn in favour of masonry

arch bridges from the above analysis:

(a) They do not require movement joints.
(b) They have proven structural reliability over more than 150 years.
(c) They can be constructed from materials that are inherently durable.
(d) The live load stresses are distributed around the structure, and not concentrated at a point, or a small number of points.
(e) They are inherently stable.

And against masonry arch bridges:

(f) They require complex analysis.
(g) They are very heavy.
(h) They can only be used for relatively short spans or for a series of short spans.

The arch is one of the oldest bridge forms and is often ignored by today's bridge designers on the grounds that the initial construction time, and therefore cost, rule in favour of beam bridge solutions. It is the authors' opinion that the inherent durability of the arch form would often lead to lower whole-life costs for an arch structure. Arches are aesthetically more pleasing than beams and can be made not only to blend in with the landscape but sometimes to enhance it. At a time when engineers are being criticised for paying insufficient attention to the aesthetic aspects of their designs we should look again at the arch as an inherently durable and elegant solution to many design problems.

## 7 References

Melbourne, C and Wagstaff M (1993) **Load tests to collapse of three large-scale multi-span brickwork arches.** Proc of Bridge Management 2, Thomas Telford, London.

Smith N.A.F. (1993) **The Roman Bridge-builder: Some aspects of his work.** The Structural Engineer, Vol 71, No 9, 4 May 1993.

Woolfenden P.A. (1993) **Modelling the masonry arch: improving modern bridge assessment using nonlinear finite-element software package (MAFEA).** Proc of Bridge Management 2, Thomas Telford, London.

# REPORT ON SESSION FIVE

B.P. PRITCHARD

Messrs Fuchs and Dumont were unfortunately unable to attend and, as a result, there was no presentation or discussion of their interesting paper.

Mr Bier described his considerable experience of continuous welded rails on Dutch railways. Dutch policy after the war was to get rid of rail joints and now 90% of track is CWR. He mentioned the very large forces involved (700 kN per rail) and the end anchor requirements, which use 60 m of track and undergo up to 60 mm movement.

Track buckling is more critical than high tension and close spacing of the stabilising sleepers is therefore necessary. Bridges carrying track cause problems and some movement between track and bridge is accommodated by suitable ballasting. Rail expansion joints are often required. Battering of rail joints is often a problem and this can be reduced, as can be the noise, by glued-in rails.

Mr Bier was questioned about the large braking effects on track and confirmed that recent research has shown much larger loading than the Dutch code had stipulated. There was some further discussion about the UK use of shock transmission units to alleviate this braking by sharing it out among joint-separated elements.

Mr Thompson provided ample evidence of the benefits of joint-free bridge construction by describing British Rail's long standing reliance on multi-span masonry railway arches. He described their ability to absorb large expansion/contraction movements by arch deformation, also their ready ability to absorb large traction and braking forces. Undoubtedly there is a case for their continuing use by the bridge engineer, particularly with BR's experience of 120 years plus life with minimum maintenance problems.

There was some discussion about the use of simple mass concrete arches with plastic sheeting separators to provide the joints normally formed between the masonry units. The pioneering work of Bolton University was mentioned.

# PART SIX

# RETROFITTING

# 18 STUDY ON JOINTLESS BRIDGE CONSTRUCTION IN JAPAN

H. TAKANO
Engineering Department, Metropolitan Expressway Public Corporation, Tokyo, Japan

**Abstract**
The paper reviews Japanese experience with embedded and combined-type jointless methods of bridge construction. Their main characteristics are described and illustrated.
Keywords: Japanese bridges; embedded joints; combined-type jointless methods; joint types.

## 1 Introduction

In Japan, with the rapid construction of nationwide road networks after the Second World War, the number of bridges increased at an accelerated pace. Among these, those built in the earlier periods were predominantly of the simple beam type, usually used in combination with expansion devices. However, it became apparent that the expansion devices were susceptible to damage due to increasing traffic volume as well as increasing numbers of heavy vehicles, resulting in deterioration of road conditions and becoming a source of vibration and noise pollution. At the same time, frequent repair of expansion devices congested traffic. Accordingly, government authorities and organizations responsible for construction and maintenance of public roads, including the Ministry of Construction, Japan Highway Public Corporation, Metropolitan Expressway Public Corporation and Hanshin Expressway Public Corporation began making concerted efforts to develop a construction method which eliminates expansion joints by paving or partially combining bridge sections, thereby achieving a continuous and smooth road surface. This resulted in the development of various construction methods for existing short span bridges, generally referred to as jointless construction methods.

*Continuous and Integral Bridges*. Edited by B.P. Pritchard. Published in 1994
by Taylor & Francis, 2 Park Square, Milton Park, Abingdon, Oxon, OX14 4RN. ISBN: 0 419 19030 9.

Jointless methods are roughly classified as follows:

- Jointless methods
  - Embedded type joints
    - Expansion absorbing type
    - Expansion dispersing type
    - Expansion inducing type
  - Combined type jointless methods
    - Main beam connection method
    - Floor slab connection method
    - Crossbeam connection method

To develop these technologies into systematic technical standards, the Committee for Study and Research on Jointless Construction Methods for Existing Bridges (Chairman: Minoru Fujiwara, Director, Bridge Structure Department, Public Works Research Institute, Ministry of Construction) was organized and a research project launched.

## 2 Embedded-type joints

There are many types of embedded joints which have been used for actual bridge construction since 1982. Following experimental installation in various projects, these joints were introduced throughout the country. The existing technology of embedded type joint, however, has not provided a perfect construction method because it is incapable of avoiding the occurrence of cracks or ruts. Nevertheless, the technological reliability has significantly improved since it sufficiently reduces damage to the joints and the adjoining area.

This method has been used in a number of bridge constructions and now is considered to have become practical. The embedded-type joints need not be permanent structure and can be installed when roads are repaved. Thus, the embedded-type joints need not be permanent structure and can be installed when roads are repaved. Thus, the embedded-type joints are sufficiently useful if they have durability similar to that of road pavement. Minor damage to the joints will not be considered serious if it causes no problems to vehicular movement. The embedded-type joints, particularly, offer such advantages as smooth driving and easy maintenance when the surface of embedded joint area is connected completely level with the adjoining road surface.

Based on such evaluation, many types of embedded joints became technically and commercially feasible for use in actual projects under certain conditions. Demand increasingly grew for government organizations responsible for road administration to compile embedded joint technologies into standard manuals. Instructions, tentatively named "Embedded-Type Joint Design and Construction Manual (proposed)," have been prepared.

In developing these manuals, a questionnaire survey was conducted among related organizations to identify how embedded-type joints were being used in actual road projects. Results showed that the number of lanes employing such joints as of the survey had reached approximately 3,000. The embedded-type joints are classified mainly into three types according to their function: expansion absorbing, expansion dispersing and expansion inducing types. The expansion absorbing type is adapted easily to absorb expansion and contraction by using soft material for the entire pavement. This type of pavement, however, is vulnerable to ruts worn by heavy traffic.

Concerning the expansion dispersing type, soft pavement material is not used, but a sheet is laid under the pavement to disperse expansion and contraction. This type prevents ruts worn by heavy traffic while the pavement adjacent to the embedded type joint is liable to crack. In short, the expansion absorbing and dispersing types are contradictory in structural characteristics.

The expansion inducing type is an embedded-type joint designed to induce expansion and contraction to a cut-off joint specifically created in the pavement. This structure, therefore, limits cracking to such specific cut-off joint area in the pavement.

The embedded-type joint is an expansion device to absorb deformation that occurs in the joint structure by utilizing the pavement's material properties and the embedded simple joint mechanism. Table 1 and Figure 1 indicate features and conceptual diagram of major embedded type joints. It should be noted that embedded type joints, due to differences in physical properties and

Table 1 Major features of embedded-type joints

| Type | Features |
|---|---|
| Expansion absorbing type | - Soft pavement materials are used, and the deformation of such materials absorbs the expansion and contraction or revolution at the extreme end of main beam.<br>- Difference in level is likely to occur between the embedded joint area and other areas because these two pavements possess different physical properties. |
| Expansion dispersing type | - Basically, the pavement structure's deformation is utilized. A sheet which is called a Shear Layer is placed between the pavement structure and floor slab. Expansion and contraction are dispersed to the entire pavement structure by the shear layer's shearing deformation. |
| Expansion inducing type | - Deformation is induced to a cut off-joint provided in the pavement structure.<br>- This method is suitable for small bridge construction. When deformation grows, however, the cut off-joint becomes a weak point and the cut off-joint area is susceptible to damage. |

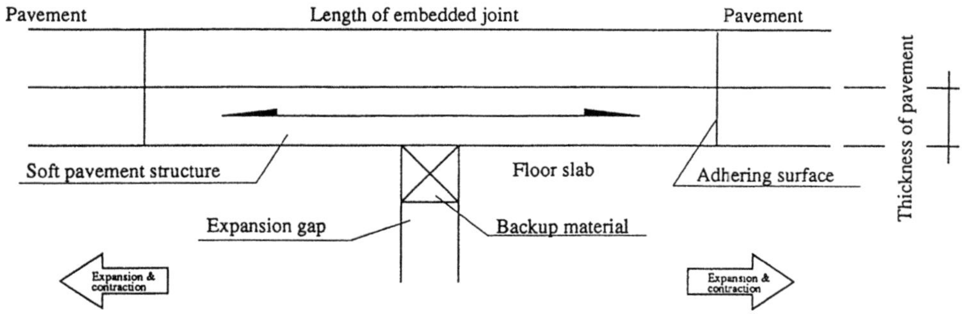


(Expansion-absorbing type)

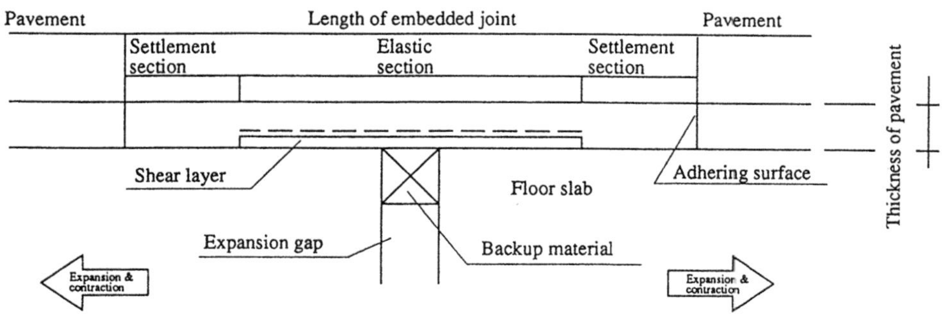


(Expansion-absorbing type)

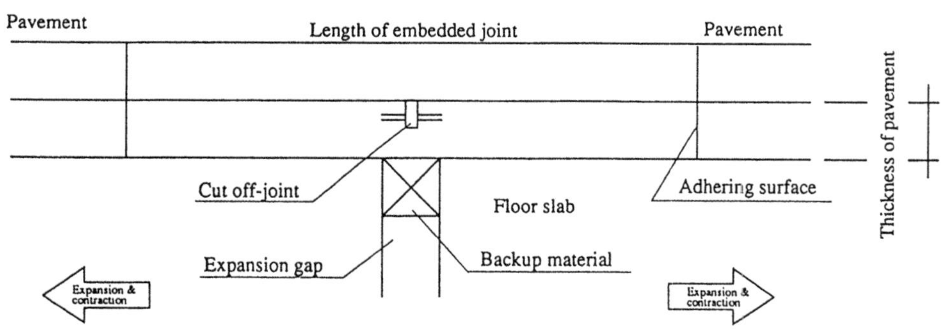


(Expansion-induction type)

**Figure 1 Conceptual diagram of major embedded-type joints**

pavement configuration, are affected by various conditions in varying degree, such as traffic conditions, type of bridge structure, maintenance requirements, and frequency of traffic congestion.

## 3 Combined type jointless method

The combined-type jointless method joins together the members of adjacent beams to transmit the temperature-induced displacement toward the bridge axis to the adjacent beams. The beams connected by combined-type joints behave in the same way as the continuous beams do when an earthquake occurs. The beam's behaviour in response to live load shows various patterns ranging from behaviour close to the continuous beam to one close to the simple beam. The beams connected by embedded-type joints behave in the same manner as the original simple beams after completion of jointless processes, and the number of spans are limited when the combined-type jointless method is applied. A combined-type jointless method should be selected according to bridge types. Table 2 indicates the combined-type jointless methods and applicable bridge type.

Application of the combined-type jointless method accompanies structural modification, which results in a larger construction scale as compared with the embedded-type joint method. The number of constructions employing combined type jointless methods is not large.

Figure 2 shows a basic classification of combined-type jointless methods.

Table 2 Combined-type jointless methods and bridge types

| Methods | Bridge type |
|---|---|
| Main beam connection method | Steel beam |
| Floor slab connection method | |
| Crossbeam connection method | PC beam |

## Acknowledgement

The author deeply appreciates Mr M Ito, Professor of Saitama University and Professor Emeritus of University of Tokyo, for giving him an opportunity to present this paper of the colloquium.

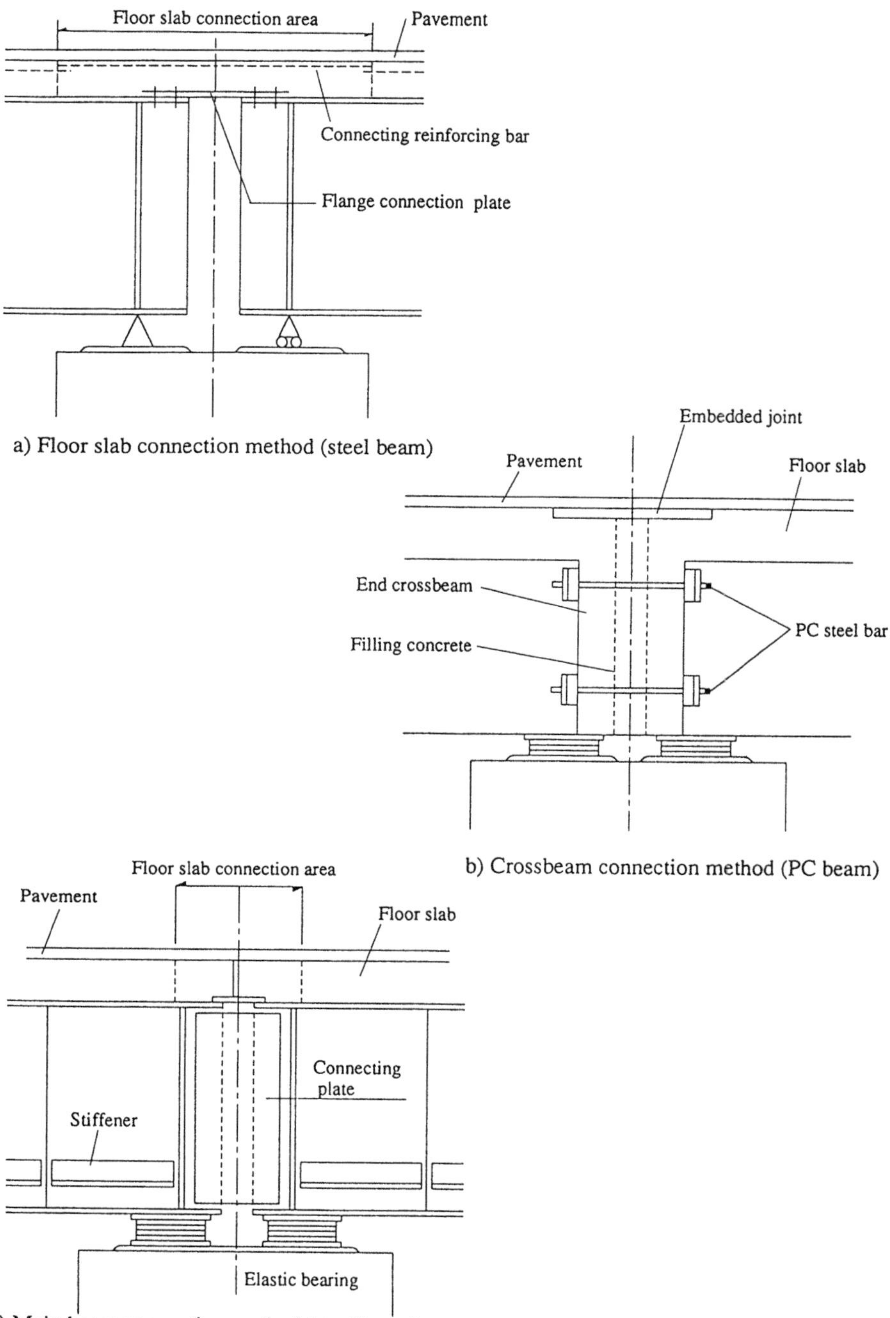


a) Floor slab connection method (steel beam)

b) Crossbeam connection method (PC beam)

c) Main beam connection method (steel beam)

**Figure 2 Basic classification of combined-type jointless methods**

# 19 SEMI-INTEGRAL BRIDGES*: A CONCEPT WHOSE TIME HAS COME?

M.P. BURKE, Jr
Burgess & Niple Ltd, Columbus, Ohio, USA

Abstract
The State of Ohio has conceived and is developing a semi-integral bridge concept that has enabled it to extend the application range of jointless bridge decks, even to bridges not normally considered for this type of construction. Although this concept has a wide application range, it possesses a number of unusual characteristics that must be recognized and be provided for if semi-integral bridges are to satisfy all functional requirements. A discussion of these characteristics is the primary focus of this paper.

## 1 Introduction

The State of Ohio has been building continuous concrete slab bridges with flexible integral abutments for 6 decades. It has been building continuous steel beam and girder bridges with flexible integral abutments for 3 decades. Although this type of construction is now routine in Ohio for most bridges, there were a number of exceptions for such applications. For example, bridges skewed greater than 30 degrees, continuous bridges longer than 300 feet (91.4 m), bridges with wall-type abutments and bridges with stub-type abutments on rigid foundations were routinely provided with movable deck joints at abutments. More recently, however, Ohio has conceived and is developing a semi-integral abutment concept that has enabled it to extend the application range of bridges with jointless decks to most moderate size bridges, even to those applications with exceptional characteristics (Figure 1). This semi-integral abutment concept is also now being used to retrofit existing end-jointed continuous bridges.

---

*Although various types and sizes of integral or jointless bridges have been constructed for centuries, the terms "integral bridge" and "semi-integral bridge" used in this paper refer to the typical small single and multiple span continuous bridges being used to bridge small streams and roadway crossings. They generally constitute the bulk of structures being constructed today.

*Continuous and Integral Bridges*. Edited by B.P. Pritchard. Published in 1994
by Taylor & Francis, 2 Park Square, Milton Park, Abingdon, Oxon, OX14 4RN. ISBN: 0 419 19030 9.

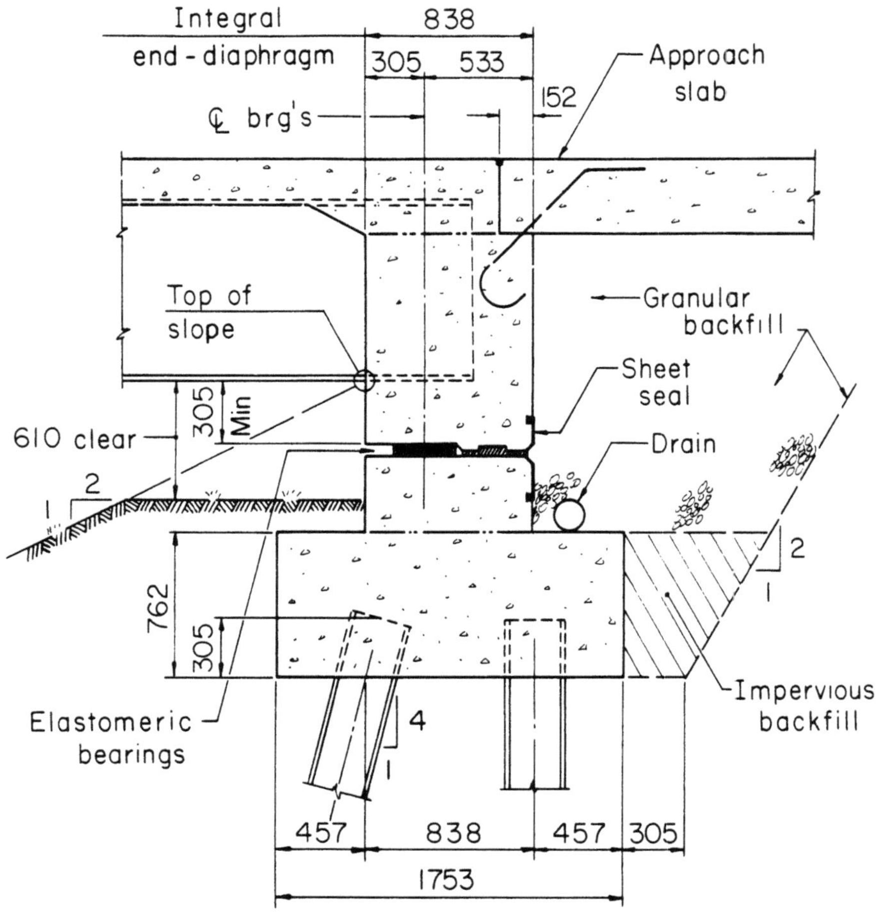


Fig. 1. A semi-integral abutment concept for single and multiple span continuous bridges.

Although the semi-integral design concept described in this paper has expanded the application range for bridges without deck joints, it possesses a number of unusual characteristics that must be recognized and be provided for. Otherwise, application of this type of design may result in bridges that do not satisfy all of their functional requirements. A discussion of these characteristics is the primary focus of this paper.

## 2 Basic features

The basic features of the semi-integral abutment concept illustrated in

Figure 1 include the absence of a bridge deck joint; a superstructure that moves longitudinally on elastomeric bearings almost independent of rigid abutment foundations; abutment members, including piling that can be designed to operate well within the usual allowable stress limits; a reduced area of structure exposed to passive pressure and pressure which is less eccentric with respect to the neutral axis of the superstructure; and abutment and end-diaphragm configurations which are simple to design, simple to reinforce, and relatively simple to construct.

Notice, however, that this design does not eliminate the need for movable joints. In fact, it doubles their number since in addition to a movable joint at the level of the bridge seat, another movable joint is needed between the approach slab and approach pavement. Yet while doubling their number, it has reduced their vulnerability to substantial maintenance. If these joints fail to function as desired, their failure will not have the damaging consequences that has come to be associated with malfunctioning deck joints and joint sealing systems.

The basic features of this design concept have engendered a number of other structural problems which must be considered and resolved for each semi-integral bridge application. These include the control of behavior at the structure-backfill interface, design of buried sealing systems for movable joints, and placement of concrete between surfaces where slight differential movements are probable.

## 3 Superstructure restraint

Of all the characteristics of the semi-integral bridge concept described in this paper, the longitudinal, lateral, and vertical restraint of the superstructure are the most important. This type of structure should not be considered for design unless its designer is familiar with these characteristics and makes appropriate design provisions to account for them.

### 3.1 Longitudinal restraint

As can be surmised by viewing the semi-integral abutment design shown in Figure 1, the superstructure which is supported on movable elastomeric bearings moves almost independent of the abutments. That is the reason why this design concept is adaptable to bridges with various types of rigid abutments. For bridges without fixed piers, it receives its longitudinal restraint almost exclusively from sources not normally utilized in bridge design for this purpose. Longitudinal restraint comes from approach slab-subbase friction, shearing resistance of elastomeric bearings, and the compressive resistance of structure backfill. However, during cold weather, after the superstructure contracts away from the abutments (and away from the backfill), only the shearing resistance of half of the bearings and the frictional resistance of the approach slabs moving relative to the subbase will be immediately available to restrain the superstructure against externally applied longitudinal forces. For this reason, it would be desirable if the granular backfill at abutments could be placed and consolidated during cold weather, or at night during hot weather so that the backfill could initially contribute more restraint to supplement that of the approach slabs and bearings for resisting longitudinal forces.

Providing turn-back wingwalls cantilevered from the superstructure in lieu of straight wingwalls would provide additional longitudinal restraint by mobilizing the resistance of backfill-wingwall friction, or for wingwalls with irregular surfaces, the shearing resistance of the backfill. For longer multiple span structures, attaching the superstructure to a free standing pier would be another way of providing additional resistance to longitudinal forces.

Generally, longitudinal resistance provided by approach slabs and bearings should be sufficient to satisfy specification requirements respecting the resistance to longitudinal forces. For moderate earthquake forces, the resistance provided by the consolidated backfill should provide the additional longitudinal restraint needed for moderate length bridges, even during cold weather. For longer bridges, anchorage to piers can provide the extra longitudinal restraint needed, even for large longitudinal forces.

### 3.2 Lateral restraint

Figure 2 shows an elevation view of a typical semi-integral abutment. Notice that the superstructure of such structures is separated from the abutment by an essentially horizontal movable joint at the bridge seat and vertical movable joints between the superstructure and transverse wingwalls. The horizontal bridge seat joint is shown in Figure 1. The vertical wingwall joints are similar except that only fillers and sealers are provided for unskewed bridges. Consequently, the abutments of this semi-integral bridge concept function essentially as longitudinal guides for the superstructure.

Vertically, the superstructure is supported by elastomeric bearings in the bridge seat joint. Laterally, the superstructure is supported by the interaction of the superstructure, approach slab, and backfill; to some extent by the compressive resistance of the filler in the lateral joints; and to some extent by the shearing resistance of the elastomeric bearings in the bridge seat joint. For applications where substantial lateral resistance is necessary (such as skewed structures [see below], or structures exposed to stream flow pressure or earthquake forces) guide bearings are necessary and their use in the wingwall joints or elsewhere between beams is recommended.

For superelevated bridges where the bridge seat joint (Figure 2) is sloped parallel to the deck surface and where the elastomeric bearings are also sloped, lateral guide bearings are necessary to provide the support needed to resist the lateral component of the superstructure reaction. Otherwise, support bearings must be set on level bridge seat surfaces.

When considering the application situations that have to be contended with in the design of semi-integral bridges, the use of guide bearings should be considered routine for most if not all semi-integral bridge designs.

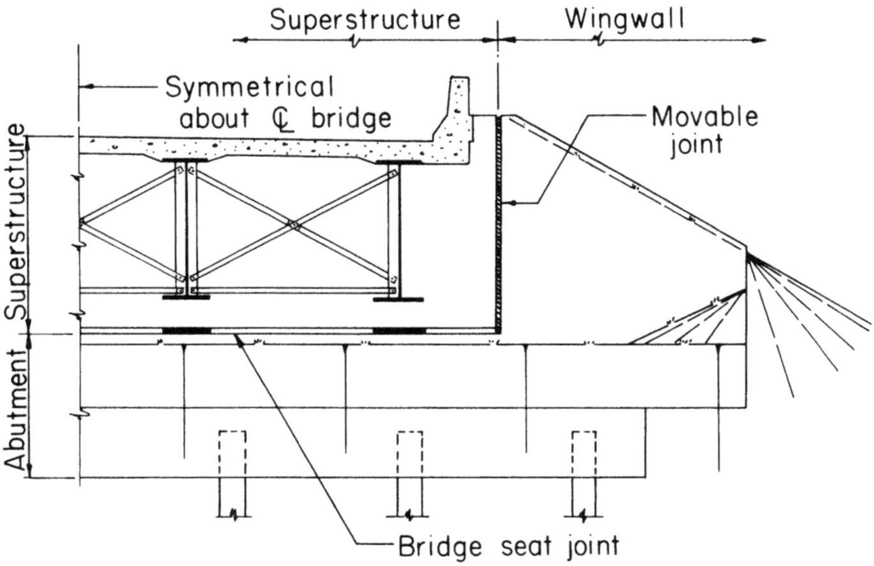


Fig. 2. Semi-integral abutment with transverse wingwalls.

### 3.3 Rotational restraint

Based on the analysis given below, superstructures of some skewed semi-integral bridges will, unless restrained by guide bearings, tend to rotate in a horizontal plane. This tendency will be greater for bridges with greater skews. Horizontal rotation will initiate sooner for longer bridges.

As superstructures of semi-integral bridges expand in response to rising ambient temperatures, superstructure elongation ($\Delta L$) will be resisted by backfill being compressed at abutments (Figure 3). Force is required to compress backfill and this same force will restrain superstructure elongation by inducing compressive stresses in the superstructure. When considering the relative compressibility of backfill and a reinforced concrete superstructure, even thoroughly consolidated granular backfill, it should be clear that almost all of the expected superstructure elongation will occur as compression of backfill. Only a slight amount of compression will occur in the superstructure as evidence by a slight reduction in the amount of superstructure elongation that would have been evident had the elongation been unresisted. These compressive stresses are shown summarized in Figure 4 as the resultant longitudinal superstructure compressive force $P_p\sec\theta$. The centralized location of this resultant force is based on the assumption that structure backfill is homogeneous, and that it would be uniformly compressed

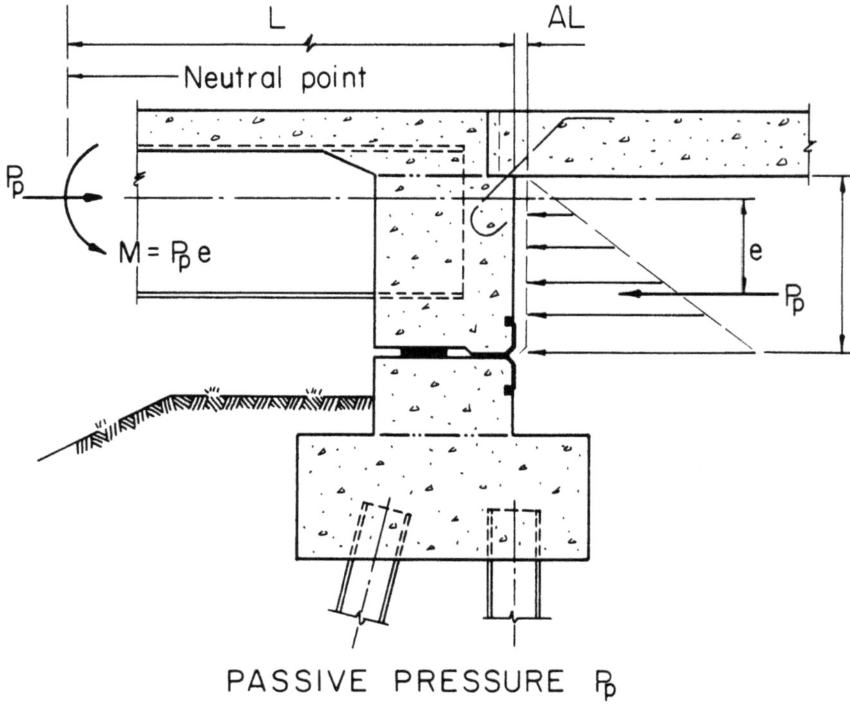


Fig. 3. Passive pressure development for semi-integral bridges.

throughout the width of the superstructure. The components of this resultant force against the backfill are the normal force due to passive pressure $P_p$ and the lateral force $P_p\sec \theta \sin \theta$, or in simpler terms, $P_p\tan \theta$. If lateral guide bearings for the superstructure are not provided and the force $P_p\tan \theta$ is not adequately resisted at the structural backfill interface (by friction of backfill on superstructure end-diaphragms [$P_p\tan \delta$] or by the shearing resistance of backfill [$P_p\tan \emptyset$]), differential movement at the structure-backfill interface will commence. When considering the shearing resistance of backfill ($P_p\tan \emptyset$) or the frictional resistance of backfill on smooth concrete surfaces ($P_p\tan \delta$), usually the latter force will be found to be the smaller of the two and it will govern behavior at the structure-backfill interface.

Since the external forces act "on" both ends of the superstructure of a semi-integral bridge (Figure 4), the eccentric longitudinal force components $P_p$ will tend to rotate the superstructure towards the acute corners of the structure, or for the skew shown in Figure 4, in a clockwise direction. The lateral force components on the other hand ($P_p\tan \emptyset$ or $P_p\tan \delta$) will tend to resist this rotation.

Using the shearing resistance of an idealized granular backfill and the frictional resistance of backfill on the backfill-structure interface surfaces, it can be shown that superstructures of semi-integral bridges skewed greater than

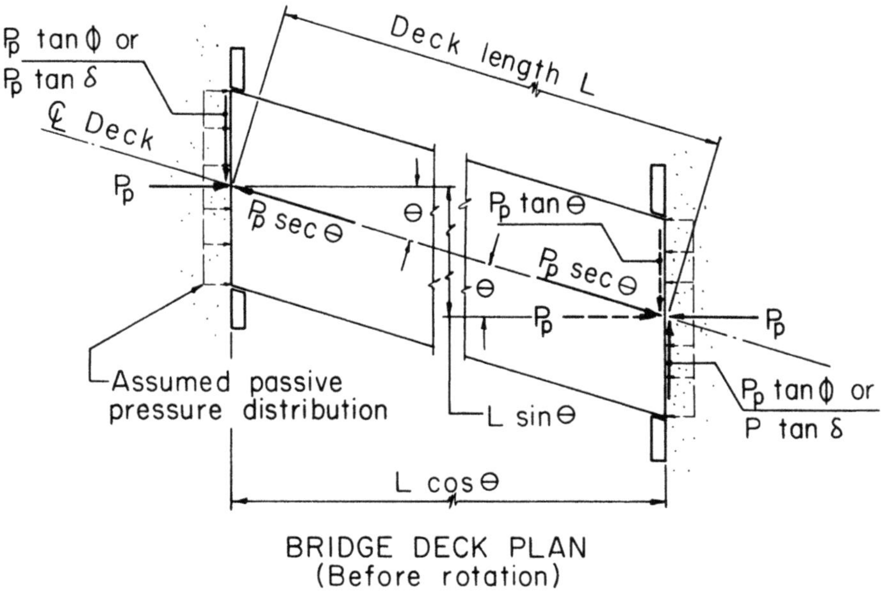


Fig. 4. The elongation and rotation of semi-integral bridge superstructures.

about 15 degrees will be unstable unless they are provided with guide bearings at both abutments.

With respect to Figure 4, the above statement can be justified by a short computation as follows:

| Where: | $L$ | = | Bridge deck length |
|---|---|---|---|
| | $\theta$ | = | Bridge skew angle |
| | $P_p$ | = | Total passive pressure |
| | FS | = | Factor of safety |
| | $\phi$ | = | Angle of internal friction of backfill |
| | $\delta$ | = | Angle of structure-backfill interface friction |

For the superstructure of a skewed bridge to be stable, the force couple tending to resist rotation ($P_p \tan \delta L \cos \theta$) must be equal to or greater than the force couple tending to cause rotation ($P_p L \sin \theta$), or;

$$P_p L \sin \theta \leq P_p \tan \delta L \cos \theta \tag{1}$$

Providing a factor of safety against rotation;

$$P_p L \sin \theta \leq P_p \tan \delta L \cos \theta / FS \tag{2}$$

Since the weight of attached approach slabs and slab-subbase friction will tend to resist movement, a safety factor of 1.5 seems sufficient for this situation. Inserting this factor in equation (2) and simplifying;

$$\sin \theta \leq \tan \delta \cos \theta / 1.5$$
$$\tan \theta \leq \tan \delta / 1.5$$
$$\theta \leq \text{arc tan} (\tan \delta / 1.5) \qquad (3)$$

Assuming that the angle of friction at the structure-backfill interface ($\delta$) is 22 degrees (see page 7.2-63 NAVFAC [1982] for granular backfill on a smooth concrete surface), equation (3) suggests that the bridge skew angle $\theta$ must be equal to or less than 15 degrees to be stable. For greater skews, it is likely that rotation will be initiated unless guide bearings are provided at both abutments to resist the forces inducing such movement.

Other observations can be made with respect to equations (1) and (3). Although equation (1) indicates that some level of passive pressure must be generated to cause rotation; equation (3) indicates that the skew angle at which rotation will be initiated is independent of both passive pressure and bridge length and directly related to structure-backfill interface friction.

The magnitude of guide bearing reactions is another indication of the potential for superstructure rotation. Since most of the thermal movement of a superstructure will be parallel to the longitudinal axis of a bridge, guide bearings should be placed parallel to this axis. Then based on the lateral force components shown in Figure 4, the guide bearing reaction, which would be normal to this axis, is given by $P_p \tan \theta \cos \theta$, or in a simpler form, $P_p \sin \theta$. Based on this relationship, the required capacity of guide bearings for a structure skewed 30 degrees is equal to $0.5P_p$, or one-half of the total passive pressure. For a 45 degree skew, the required capacity equals more than 70 percent of the total passive pressure. Consequently it is clearly evident that neither the frictional resistance ($P_p \tan \delta$) nor the shearing resistance ($P_p \tan \phi$) of the backfill can resist forces of this magnitude. Consequently, guide bearings should be provided for structures with large skews if a stable superstructure and a fully functional bridge is to be provided.

### 3.4 Vertical restraint

Due to their jointless construction, many types of integral bridges are buoyant when they become submerged. This is true for many I-beam bridges and some spread-box beam bridges.

The weight of diaphragms at abutments provide some resistance to uplift. But generally, some positive design provisions must be made to ensure that integral bridges have a reasonable factor of safety against flotation. I-beam webs can be pierced near top flanges by 3-inch diameter holes spaced uniformly throughout the beam length; the space between spread boxes can also be vented by placing 3-inch minimum diameter horizontal vent ducts near the top flange of all beams. These ducts should pass completely through the beams from one web to the other and they should be placed in concrete diaphragms or be completely encased in concrete to prevent floodwaters from entering beam voids; counterweights could be used but their weight must be taken into account during beam design; uplift restraints could be provided at pier bearings; or some piers can be integrally constructed with the superstructure to add sufficient uplift restraint to counteract buoyancy.

In lieu of vent holes, added weight, uplift restraints, or integral pier construction, the use of the most buoyant structures should be restricted to those bridge sites where the highest floodwater levels are well below the superstructure.

## 4 Design aspects

### 4.1 Movable joints

As has been mentioned, the semi-integral bridge with its attached approach slabs has eliminated bridge deck end-joints. In their place, it has incorporated two other joint types, a movable joint at the level of the bridge bearings, and one at the pavement end of the approach slab. Although doubling their number, the design has minimized their significance. Less than desirable performance for either of these joints will not have the significantly adverse consequences that has come to be expected with the failure of bridge deck joints. The bridge seat joint and cycle control joint are the two joints that have been provided to accommodate the movement of semi-integral bridges. In addition, one must mention that rigid approach pavements also must be provided with effective pressure relief joints to guard semi-integral bridges from uncontrollable approach pavement growth.

### 4.11 Bridge seat joints

In the design shown in Figure 1, the troublesome bridge deck joint has been eliminated. However, in its place, a movable joint has been introduced at the level of the bridge seat. Corrosion resistant elastomeric bearings are provided so that the superstructure can move longitudinally almost independent of the rigid abutments.

The movable bridge seat joint must be provided with a durable elastomeric seal since it is buried in the backfill and consequently is not accessible for repair or replacement. Otherwise its most important characteristic is its ability to prevent backfill from being forced into the joint by compressed backfill. It would be desirable but not absolutely necessary for the seal to be watertight. It must also permit unrestrained differential movement between the abutment and superstructure, even for bridges with large skews, and it must retain these characteristics for many years without the need for repair or replacement.

Although the bridge seat joint seal is an important aspect of the semi-integral bridge design, Ohio has yet to adopt a design that appears to fulfill all of the necessary functional and durability characteristics. A number of trial designs have been developed and used. Initially, standard compression seals were employed. Then it became apparent that a reinforced elastomeric sheet type seal was more functionally suitable for both square and skewed applications. The sheet seal now being used is a 3/32 inch thick nylon reinforced neoprene. It is attached to the bridge by various means including elastomeric anchor rods in formed recesses, steel clamp bars with expansion anchors, washers and masonry nails, or bonding adhesives. It remains to be seen which one or more of these attachment methods will be adopted and perfected by Ohio for this critical joint.

### 4.12 Cycle control joints

Semi-integral bridges with attached approach slabs lengthen and shorten in response to temperature and moisture changes. Consequently, for such structures, the boundary between approach slabs and approach pavement should be provided with cycle control joints to facilitate such movement. Otherwise, longitudinal cycling of both structure and approach slabs can damage both flexible and rigid approach pavements.

At present, standard pavement expansion joints with compressive fillers are being provided for shorter semi-integral bridges. Longer bridges are being provided with pavement pressure relief joints (wide joints filled with asphalt concrete).

For semi-integral bridges built adjacent to rigid pavement, it is imperative that they also be protected from pressures due to the restrained growth of approach pavement. Effective pressure relief joints should be provided for all semi-integral bridges, even the shortest bridges. Consequently, for bridges adjacent to rigid approach pavement, two types of pavement joints are required; one to facilitate the cyclic movement of the bridge and the other to protect the structure and cycle control joints from the effects of pavement growth. Designs by four Transportation Departments are illustrated by Burke (1987). As noted in that reference, all designs presently being used have their limitations.

For longer integral bridges, Ohio is using wide pressure relief joints to serve both purposes. Since integral bridges and semi-integral bridges are such new conceptions, much additional development is needed if the approach pavement joints adjacent to such structures are to provide all of the necessary attributes that these joints must have to satisfy structure requirements without continuous maintenance.

### 4.2 Backfill

Backfill for semi-integral bridges should not be considered a nuisance that has to be contended with as is the case with the fully integral bridge on flexible abutments. Instead, backfill should be recognized as an integral part and an important part of the semi-integral bridge concept. As in the case of a retaining wall supported by spread footings on soil, the wall when properly designed will interact compositely with the soil and be adequately supported by the soil both vertically and laterally. In a similar fashion, the superstructure and backfill of semi-integral bridges form a partially composite interactive structure. In this context, the backfill performs multiple functions. While rigid abutments provide both vertical and lateral support for the superstructure, the backfill supplements this support by providing vertical support for approach slabs and both longitudinal and lateral support for the superstructure. And the ultimate success or failure of the semi-integral bridge concept will depend to a great extent upon methods and procedures that are developed by the bridge engineering profession to enhance the interaction between the superstructure and backfill.

Since backfill is such an integral part of a semi-integral bridge concept, every effort should be made to ensure that it is properly selected, constructed, and maintained.

Backfill should be selected and designed to have characteristics suitable for superstructure-backfill interaction; it should be of a composition that protects it from erosion and it should be protected from above by full-roadway width approach slabs. For bridges with confined drainage (raised curbs, barriers, parapets, etc.), approach slabs must be provided with curbs with a height of at least 6 inches or more to confine roadway drainage and conduct it along bridge approaches and away from the backfill. Approach roadway curb inlets should be considered and be provided if necessary to ensure effective drainage control.

An effective subdrainage system should also be provided in the backfill above impervious embankments to ensure that the retention of subsurface water is minimized.

Provisions also should be made to intercept subsurface approach roadway drainage and discharge it away from the abutment backfill. Granular subbases should be provided with efficient lateral drains to discharge subbase drainage laterally to embankment side slopes. Roadway underdrains must be terminated beyond the bridge approach slabs and be provided with lateral drains to embankment side slopes. Otherwise, underdrain accumulations should be conducted in closed conduits longitudinally through the backfill and abutments.

Finally, bridge maintenance engineers should become familiar with semi-integral bridge characteristics so that they can properly appreciate the importance of backfill superstructure interaction and provide the corrective maintenance that such structures must have if they are to provide the service life that their design anticipated.

## 5 Construction aspects

Unlike their jointed bridge counterparts, semi-integral bridges have concrete and backfill placement peculiarities that must be recognized and be provided for during design. Placement of end-diaphragms and approach slab concrete must be limited to the extent that the thermal movement of the superstructure during and shortly after concrete placement is minimized to preclude damaging freshly setting concrete. This can be done by limiting concrete placement to those days when rapid temperature changes are not expected and to periods during the day when superstructure movements are smallest, generally shortly after the ambient temperature approaches, reaches, and departs from the day's peak temperature. Backfill at abutments must be placed and compacted simultaneously at both abutments to avoid displacing the superstructure.

## 6 Summary

The first bridge with semi-integral characteristics similar to those that have been described in this paper was constructed in 1978. This bridge services Ohio's Route 555 and spans the Muskingum River at Zanesville, Ohio. It is a 540 foot (164.6 m) long unskewed three span girder structure. It employs approach slabs and turn-back wingwalls to engage or embrace the backfill. Since then a number of similar shorter structures have been constructed. The concept has been used most often to retrofit existing end-jointed bridges. Presently, a number of other semi-integral bridges are being planned for both new and retrofit applications, some with significant skews. As of this writing, skewed semi-integral bridges with guide bearings have not yet been constructed.

The response of local maintenance engineers to these bridges has been good. It is primarily through their continual urging that many of these bridges were built.

For new structures, the main emphasis of the Ohio Department of Transportation (ODOT) is on the construction of fully integral bridges with flexible abutments. However, for those applications where rigid abutments are necessary, the semi-integral bridge concept is now being adapted and used with increasing regularity. The actual performance of these bridges throughout the next several years will influence its further development and ultimately its suitability for further applications.

## 7 References

Burke, M. P., Jr., (1987) **Bridge Approach Pavements, Integral Bridges, and Cycle Control Joints**, Transportation Research Record 1113, Transportation Research Board, National Research Council, Washington, DC.

NAVFAC, (1982) **Foundations and Earth Structures**, Design Manual 7.2, Department of the Navy, Naval Facilities Engineering Command, Alexandria, VA.

# REPORT ON SESSION SIX

D.W. QUINION

The title 'retrofitting' was really a misnomer for two papers on experience in Japan and the USA. Mr Takano amplified on the experience gained in Japan with various types of embedded-type jointings and combined-type jointless methods. The former are extensively used on the basis of predicted performance and the accumulating experience of test programmes. The joint variants allow for the different pavement characteristics. The jointless method is a treatment providing structural continuity. It was commented that early experience in the UK of such embedded-type joints is very similar to that described. There had been welding difficulties with steel laminated bearings both in Japan and the UK.

Mr Burke described how Ohio have a long experience of building semi-integral bridges and this type of bridge is now being used frequently in neighbouring states learning from that experience. He asserted that elastomeric bearings are now regarded as very reliable and no cause for concern. He described the usual ways in which they are installed which indicated that they could be replaced if necessary. There was discussion about the sealing of the gap between abutment and the deck. Suggestions were made that an alternative way using an access gallery could provide easy access to the bearing and the underside of the joint at pavement level whilst providing protection to the bearing against loose materials and water. Whilst it drew on a common German practice regarding access galleries Mr Burke considered it to be unnecessary as well as expensive.

Mr Burke described 'run-on-slabs' as being very reliable as they were conservatively reinforced against any loading conditions. They had never been called into question. There was also discussion on the needs for and provision of restraint against horizontal rotation of strongly skewed bridge decks. Mr Burke had emphasized the need for guide bearings to prevent this. UK members commented that post-tensioning of skew bridge decks bearing on slender piers can result in a redistribution of stress and deflections unless the deck is restrained against rotation.

PART SEVEN

# PRECAST BEAM DECK CONTINUITY

# 20 CONTINUITY OF PRECAST PRETENSIONED CONCRETE BEAMS BY EMBEDMENT IN INTEGRAL CROSSHEADS

B.P. PRITCHARD
Consultant to W.S. Atkins Consultants Ltd and
Colebrand Ltd, Haywards Heath, UK

**Abstract**
The relative advantages of the five types of continuity methods which can be used for precast, pretensioned bridge beams are discussed. These are: wide and narrow in-situ integral crossheads; integral crosshead cast in two stages; continuous separated deck slab; and tied deck slab. It then describes a survey of existing bridges constructed by these methods which was carried out for the Transport Research Laboratory.
Keywords: Precast pretensioned concrete beams, beam continuity by embedment, deck slab continuity, deck slab hinging.

## 1 Introduction

The advantages of precast concrete beam deck construction of ease and speed of erection, elimination of obstructive falsework, standardisation and factory qualities brought to site are well known. As a result it is one of the most favoured construction techniques for small to medium span bridge decks.

The most commonly used prestressed beams are the 'M' and 'U' series, supporting and composite with an in-situ reinforced deck slab. Multi-span bridges rarely feature continuous decks and are usually built as a sequence of simply supported spans. With bearings required under each beam, support piers are of deck width, formed either as solid leaf or columns and crosshead arrangements.

Precast concrete beam deck construction has generally been limited to simply supported structures. When multi-span, it is easier to design and construct the deck as a sequence of simply supported spans than to incur the added complexity and time penalties by adding continuity. These difficulties are often more perceived than real.

In addition, multi-span simply supported decks are usually confined to straight or near straight alignments because of difficulties in fitting straight beam elements into curved decks. Plan curvature requires varying beam lengths across the width of each span, removing most of the benefits of standardisation. Tight vertical curvature leads to excessive in-situ deck slab build up at midspan of the chorded beams, adding unnecessary deck depth and weight. As a result, this

*Continuous and Integral Bridges*. Edited by B.P. Pritchard. Published in 1994
by Taylor & Francis, 2 Park Square, Milton Park, Abingdon, Oxon, OX14 4RN. ISBN: 0 419 19030 9.

form of construction has rarely been considered for the convoluted in plan and vertical profile flyovers often found in tightly confined urban situations.

However, the emphasis is changing and continuity is now a highly desirable feature because of the reduction of expansion joints. Corrosion caused by winter de-icing salts is currently the most damaging threat to the integrity of UK bridges. Investigations for the UK Department of Transport[1] have confirmed that most bridge deck expansion joints leak and contribute more than any other bridge element to deck and substructure corrosion damage. The joints themselves also have a poor maintenance record.

## 2 Precast concrete beam deck continuity

Precast concrete decks invariably carry and act compositely with an in-situ concrete deck slab. Why not use some of the in-situ concrete construction in the pier area, where it is well suited to solving the problems of curvature and continuity besides offering further advantages in pier construction and deck drainage arrangements? Confining this construction to an area over the pier foundations gives readily available support for staging, with end embedded precast beam/in-situ slab construction in the deck areas between and provides the individual advantages of in-situ and precast construction at the localities best suited.

The Department of Transport is shortly to issue an Advice Note which will encourage the use of continuity in multi-span bridge decks in order to achieve the desirable reduction of expansion joints. With the precast concrete beam deck representing one of the most commonly used forms of deck construction, the Transport Research Laboratory recently commissioned an investigation of the ways in which continuity can be provided for composite bridge decks where the primary beams are of pretensioned precast concrete[2].

The study identified five distinct methods of building in various degrees of continuity and eliminating intermediate deck expansion joints in the UK. Some of the methods are used internationally. One of the methods also overcomes most of the urban flyover curvature problems described in the Introduction.

The classification of the methods indicated that two make only the deck slab continuous or hinged over the piers whilst the other three provide full monolithic continuity connection of the beams and slab over the piers.

All methods require in-situ concrete transverse beams at or near the piers. These beams can be arranged as twin, on either side of the pier, or singly, located over the pier. Transverse beam depth can be less than, equal to, or greater than the longitudinal beam depth. For clarity, single transverse beams located over the pier and pier bearings are designated as integral crossheads. All other transverse beams are designated as diaphragms.

### 2.1 Type 1 - Wide in-situ integral crosshead

This type of continuity detail (Fig. 1) uses precast pretensioned beams significantly shorter than the spans between support piers and is the most

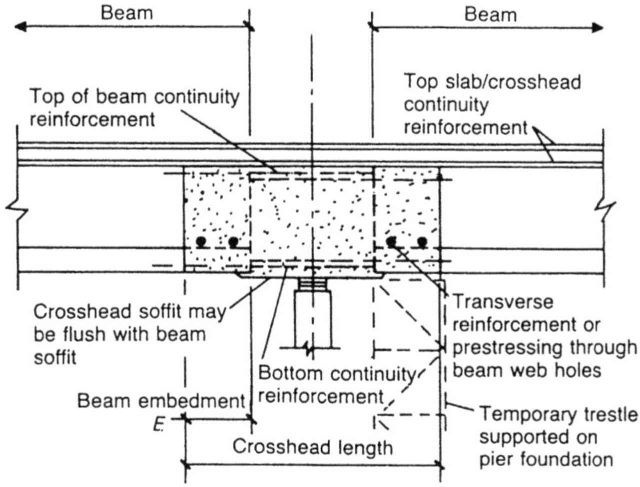


Fig. 1 Type 1 continuity connection: wide in-situ integral crosshead.

commonly used UK method. The beams, generally 'U' or 'M' type, are usually supported on temporary trestles built off the pier foundations. The wide in-situ integral crosshead over the pier is then cast between and around the beams to provide about 1 m embedment. Longitudinal continuity is accomplished by reinforcement within the continuous composite deck slab, generally supplemented by reinforcement and, sometimes, pretensioning strand ends extending from the top and bottom of the embedded beams. Transverse strength is assured by either prestressing tendons or reinforcement, some of which may pass through holes in the ends of the precast beams. The crosshead is supported on a single row of bearings set centrally on the pier.

**2.2 Type 2 - Narrow in-situ integral crosshead**

This continuity detail is used extensively in America (Fig. 2). The precast pretensioned beams are long enough to be erected on two parallel rows of temporary or permanent bearings on the pier tops. As with Type 1, the in-situ integral crosshead over the pier is then cast between and around the beams to provide about 1 m embedment. The crosshead is, however, narrower than Type 1 because of the necessarily small gap between the embedded beams. The same narrow gap makes adequate reinforcement connection difficult between the beam ends. Longitudinal hogging bending continuity is again readily established by top reinforcement within and extending well into the continuous composite deck slab. Transverse strength of the crosshead is generally provided for by reinforcement, some of which passes through holes in the ends of the precast beams.

Where twin rows of temporary bearings are used, a central row of permanent bearings located under the crosshead is brought into use by removing the temporary bearings after the crosshead concrete has gained sufficient

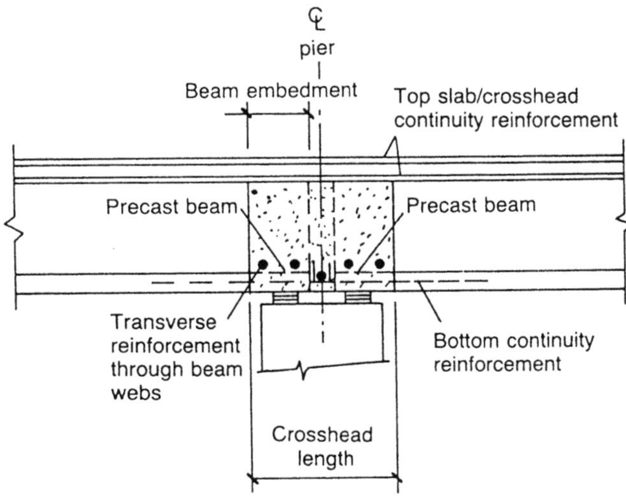


Fig. 2 Type 2 continuity connection: narrow in-situ integral crosshead.

strength. Some examples use a wide single permanent rubber bearing which acts as a seating for both beams.

### 2.3 Type 3 - Integral crosshead cast in two stages

This type of continuity detail is a variant of Types 1 or 2 where the integral crosshead is cast in two stages (Fig. 3). The crosshead is of greater depth than the main deck precast beams and the bottom section is cast first to support these beams, generally on thin mortar beddings. The second stage completes the integral crosshead in the manner described for Type 1.

### 2.4 Type 4 - Continuous separated deck slab

This continuity detail confines itself to the deck slab only, which flexes to accommodate the rotations of the simply supported deck beams erected in the conventional multi-span manner (Fig. 4). To permit this flexure, the deck slab is separated from the support beams for a length of about 1.5 m by a layer of compressible material. In-situ reinforced concrete diaphragms are located at the ends of the separated slab.

A variation of this method omits the diaphragms and is known as the continuity plate construction[3]. This was introduced in America in the fifties and tested and developed in Switzerland and Germany in the sixties. In Bavaria, Germany, over 300 multi-span precast concrete beam bridges had been built by 1981 using this Type 4 variant. The detail was reported as practical, economical and durable and is still being constructed in Germany.

### 2.5 Type 5 - Tied deck slab

This tied deck slab detail was developed for multi-span precast beam decks during the UK Standard Bridge exercise of the 1970s[4]. The bridge decks are

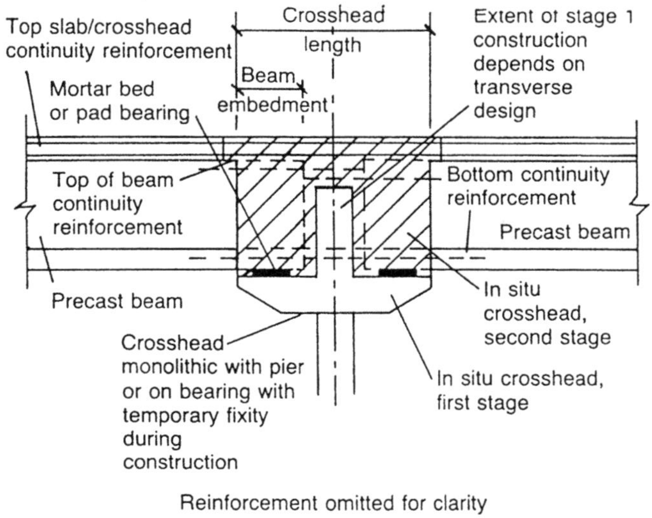


Fig. 3 Type 3 continuity connection: integral crosshead cast in two stages.

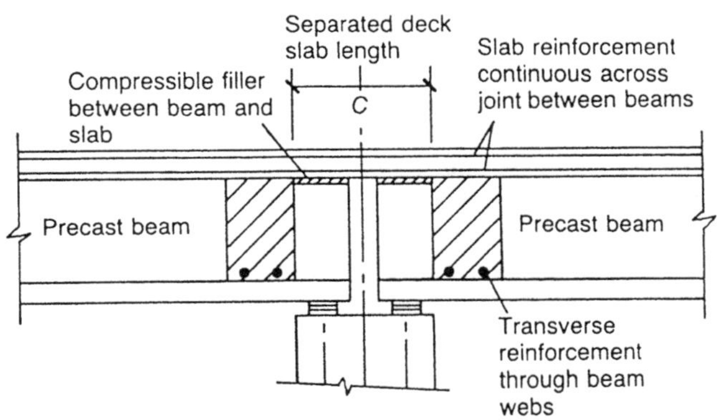


Fig. 4 Type 4 continuity connection: continuous separated deck slab.

designed and constructed in the conventional multi-span simply supported manner with slab trimmer diaphragms at the beam ends. These ends are, as with Type 4, carried on two parallel rows of bearings on the piers.

Long connecting reinforcement dowels are incorporated at the slab mid-depth to tie the slabs together over the pier, eliminating expansion movement at deck level and permitting the use of buried deck rotation joints. To accommodate this rotation, the dowels are debonded and sleeved from the surrounding slab concrete over short lengths either side of the joint. Also, the slab and trimmer beam downstands are 'necked' using compressible joint filler below and above the dowel connection.

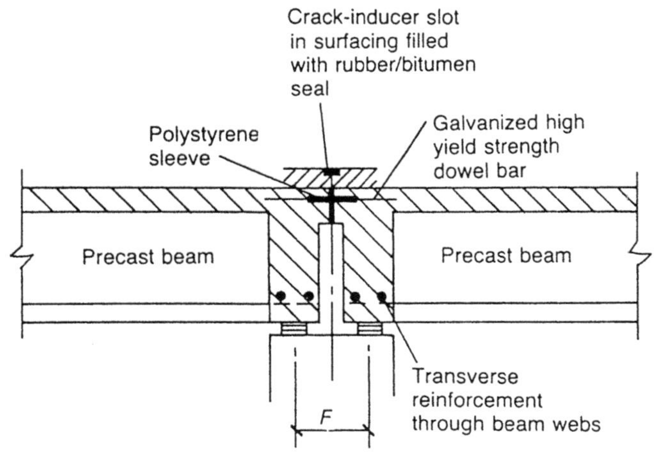


Fig. 5 Type 5 continuity connection: tied deck slab.

## 3 Choice of continuity method

If the main requirement is to eliminate troublesome intermediate deck expansion joints in multi-span decks then Types 4 and 5 offer the minimum of extra design and construction effort. The simplest procedure, Type 5, is however probably the least durable because the tied deck joint retains a rotational capacity, with a sealed notch required in the deck surfacing.

However, to gain the advantages of part deck continuity, one of the Type 1, 2 or 3 methods should be adopted. The deck bending reductions are modest but still worthwhile in terms of construction depth savings and associated savings in the lengths of flyovers and/or approach ramps. Modest, because the precast beam self-weight element remains simply supported and further continuity moments due to restrained creep and shrinkage of the beams must be considered in addition to the normal surfacing, live load and temperature loadings. Nevertheless, typical savings allow M8 beams to be used in place of simply supported M10 beam decks.

The embedment of the precast pretensioned beams into the in-situ crosshead involves restraint to the continued prestressing creep rotation of the beam ends, giving rise to sagging moments at the supports. The beam ends rotate in the opposing direction due to the later differential shrinkage set up in pouring the in-situ concrete deck slabs, giving rise to hogging moments over the supports. In general, the two moments tend to eventually cancel each other. However, there is usually a one or two month delay between beam embedment and slab pouring and the moments due to the restrained creep during that delay period must be considered alone for one of the design loading combinations.

In the following sections, the relative advantages of Types 1 and 2 methods are listed. Type 3 method, representing an amalgam of Types 1 and 2, is not particularly favoured for two reasons. Firstly, the downstand half of the crosshead is obstructive, both aesthetically and in terms of headroom. Secondly, to take advantage of a single row of bearings under the crosshead, special care must be taken during beam erection to prevent crosshead rotation and collapse, which has happened, due to unsymmetrical erection of the precast beams on each side of the crosshead.

### 3.1 The relative merits of Type 1 method

This, the most commonly used method in the UK, is relatively more complex to design and construct than any of the four other methods. Nevertheless, it offers more advantages, including the ability to accommodate plan and vertical curvature.

The key element is the wide in-situ concrete integral crosshead, with temporary beam support trestling and crosshead formwork supported off the pier foundations. The wide crosshead offers the following advantages:

(a) Precast beam deck spans are most often limited to the readily transportable length of 27.5 m defined by the Department of Transport. With 5 m wide crossheads and 1 m beam embedments, the continuous span lengths can be extended to 30.5 m. The usual end projecting reinforcement can be removed from embedded screw couplers during transport.

(b) Plan curvature can be readily accommodated by varying the width of the integral crosshead to form a plan trapezium shape. This overcomes the usual problem of the beams on the outside of the curve being longer than those on the inside and permits the use of one standard length beam. The side curvature is taken up by varying the width of the facing concrete.

(c) Vertical curvature problems can be reduced by vertically curving the top and bottom surfaces of the crosshead. This reduces the increased slab thickness at midspan required to take up the vertical curvature above the straight chorded precast beams.

(d) Drainage of surface water on multi-span viaducts ideally requires collector downpipes spaced no further apart than the pier support positions. This means lateral connecting piping between kerb gulley drains at the deck edges and downpipes within or attached to the sides of the piers. Normal simply supported composite beam span decks are so congested with precast beam ends over the piers that it is usually too difficult to pass the drainage pipes through. Resort has therefore to be made to ugly external piping passing very prominently down the deck sides or to more complex longitudinal deck drainage systems. The Type 1 continuity method uses wide in-situ integral crossheads with the precast beam ends embedded some distance from the pier. This readily allows enough room for the passage of a concealed internal lateral connecting drain using generous diameters and bends (Fig. 6).

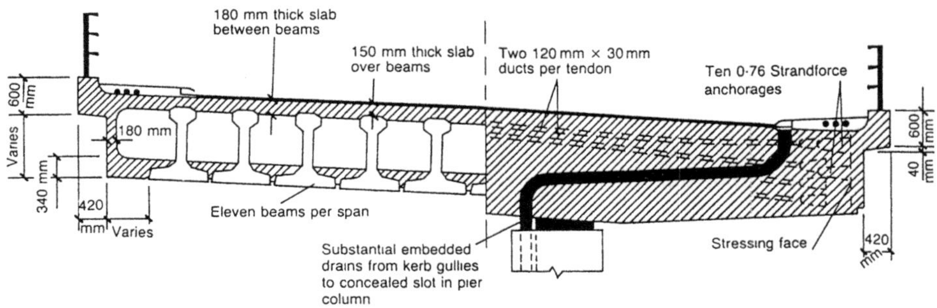


Fig. 6 Concealed drainage piping.

(e) Type 1 uses a single row of bearings. This immediately halves the numbers of bearings required for simply supported construction, although individual bearing size will marginally increase.

It also means that piers are thinner, not only because the single line of central bearings takes up less room at the pier top, but because the dead and live load moments applied to the piers by off centre pairs of bearings are removed. Significant savings in pier foundations also result.

Full width piers or crossheads are not required either. The integral crosshead in the deck can be top reinforced or prestressed laterally to allow considerable deck cantilevering outside the pier. Resulting inboard piers can offer considerable savings in ground intrusion. This in turn can allow ground level slip roads to be located under the deck edges and can reduce flyover spans at skew crossings, both particularly important in cramped urban locations. There is also a useful saving in further numbers of bearings.

The largest use of Type 1 method in the UK was during the construction of the six viaduct London to Cambridge M11 South Woodford Interchange in 1976[5]. This was followed by the construction in two further similar viaducts forming part of the South Woodford to Barking Relief Road in 1987. The crossheads were transversely prestressed, as much as anything to avoid congestion in the upper surface due to heavy transverse and longitudinal reinforcement. Tests of the crosshead were undertaken at the then Cement and Concrete Association at Wexham Springs[6]. These demonstrated the effectiveness of the shear transfer between the beams and crosshead and proved the method to be entirely successful. Nevertheless, the transverse prestressing is expensive and later constructions have resorted to the use of normal steel reinforcement rather than prestressing.

### 3.2 The relative merits of Type 2 method

This is almost the only continuity method used in America and is relatively easier to build than Type 1. Nevertheless, it cannot offer the advantages of extra span, curvature or drainage offered by Type 1. In most cases, it also does not offer the bearing and pier advantages, unless beams are seated upon temporary twin rows of bearings and loads are eventually transferred to a permanent single row of bearings set under the integral crosshead, or use is made of a wide single bearing.

Undoubtedly, the greatest advantages of Type 2 method lies in the ease of placing the precast beams directly on to the piers. Nevertheless, the narrower integral crosshead makes adequate reinforcement more difficult between beam ends, particularly for sagging moment bottom reinforcement.

A recent American report[7] has proposed that the provision of this bottom reinforcement should be optional provided that continuity moments are calculated assuming lack of positive moment continuity. It is also suggested that any cracking resulting from not providing this reinforcement is no threat to the deck integrity and will certainly eliminate any corrosion possibility. No doubt a compromise solution could be used - a limited amount of anti-crack reinforcement in the bottom of the crosshead, possibly epoxy coated to prevent corrosion.

Major research and testing of the Type 2 connection was undertaken by the Portland Cement Association in Chicago in the early sixties[8]. It covered all aspects, ranging from creep and shrinkage to horizontal shear connection. It also covered extensive testing of a half scale continuous two span deck, some 20 m long x 5 m wide. A recent UK publication[9] sets out design procedures for using Type 2 continuity for the new precast Y beams.

## 4 References

1 Wallbank, E. J. (1989) **The performance of concrete in bridges - a survey of 200 highway bridges.** London HMSO.

2 Pritchard, B. P. and Smith, A. J. (1991) **Investigation of methods of achieving continuity in composite concrete bridge decks.** TRL Contractor Report CR 247. Crowthorne.

3 Wagner and Buchting (1981) **Continuity plate construction in multi-span bridges using precast concrete beams.** Bauingenieur.

4 Department of Transport (1979) **DTp standard bridges.** London HMSO.

5 Pritchard, B. P. (1982) **Integral in-situ crossheads for continuous precast beam deck viaducts.** International Conference on short and medium span bridges. Toronto.

6 Sturrock, R. D. (1974) **Tests on model bridge beam in in-situ concrete construction.** Cement and Concrete Association.

7 Transportation Research Board (1989) **Design of precast prestressed bridge girders made continuous.** National Cooperative Highway Research Program Report 332.

8 Mattock, A. H. Kaar, P. H. Hanson, N. W. Hognestead, E. and Kriz, L. B. (1960-61) Precast concrete bridges. Seven papers in **Journal of Research and Development Laboratories.** Portland Cement Association.

9 Hambly, E. C. and Nicholson, B. A. (1991) **Prestressed beam integral bridges.** Leicester, Prestressed Concrete Association.

# 21 LOCALLY SEPARATED DECK SLAB CONTINUITY IN COMPOSITE BRIDGES

A. KUMAR
Kumar Associates, Beaconsfield, UK

Abstract
This paper considers the problems associated with jointed deck construction and commonly used remedies leading to logical evolution of ideas and concepts of locally separated deck slab continuity in composite bridges. These are further advanced to practically usable details in harmony with the overall structural behaviour of bridges. Proper application of these should result in savings to the exchequer, increased life and trouble free performance of such bridges.
Keywords: Deck slab continuity, Top slab continuity, Locally separated deck slab, Connecting slab, Continuity plate, Isolation plate, Multi-span composite bridges, Continuous composite bridges, Continuity with precast bridge beams, Continuity with steel beams, Concepts of continuity.

## 1 Background

Span by span simply supported construction of multi-span bridges using some form of precast (or steel) longitudinal beam elements has been in vogue since about 1950's. During the sixties and seventies the need for rapid expansion of the country's trunk road network combined with no falsework requirement and shorter period of deck construction, the numbers of such multi span bridges rapidly increased. Due to the public requirement of keeping roads open through the winter months, this period also witnessed an increasing use of salt on British roads.

During the seventies inspections of such bridges had started revealing that despite every effort in devising competent joint details, salt water penetration through the joints and consequential damage to deck ends and support elements, was wide spread. The pressures of rapid expansion of roads network and simplicity of design and construction are however such that the practice of jointed construction has continued largely unchallenged through the eighties to the present day.

*Continuous and Integral Bridges.* Edited by B.P. Pritchard. Published in 1994
by Taylor & Francis, 2 Park Square, Milton Park, Abingdon, Oxon, OX14 4RN. ISBN: 0 419 19030 9.

Estimating the cost implications of deterioration due to such construction is quite difficult as it involves varying combinations of repair and replacement costs, reduction in useful life of structures and traffic disruption etc. Even if the latter is excluded, preliminary estimates place the one going cost of such deterioration at around £10m/year. Although this is not a particularly large sum in the context of government's expenditure on roads, it is the author's view that such direct costs together with associated unquantifiable traffic disruption costs, would be significant and could be largely eliminated by using the concepts presented here.

## 2 Existing remedies

As mentioned earlier the problem of joints deterioration has been recognised for some 20 years in the UK and for about 30 years in the USA. During the early sixties in the USA and seventies in the UK, researches had been conducted to eliminating joints in such construction. The resulting so-called 'live load' continuity solutions in which large insitu concrete diaphragms embedding the ends of precast units are used, do eliminate the intermediate support joints and hence such deterioration. However the design process is greatly complicated due to secondary effects such as temperature, shrinkage and creep becoming important design parameters. The manufacture of precast elements and construction of the deck are also made considerably more complicated because of the need to accommodate reversals of moments in the precast and insitu concrete adjoining the intermediate support positions. This also generally negates any anticipated reduction in depth of construction of such decks. These factors in reality can lead to cost increases of the order of 25% to 50% in the cost of deck construction depending on the number of spans involved. Such increases in cost could amount to additional expenditure of up to £10m per year, if this remedy is applied to all such bridges.

The so called 'integral' bridges could also be viewed as an extension of this approach, in that the abutments are also integrally connected to the bridge deck with no provision for movements except by sliding of abutments on foundations. It is however doubtful whether the additional expense and empiricism in design is justifiable against the cost saving of an end expansion joint as well as the likelihood of disintegration of soil wedges behind abutments and inderterminate state of stress in the structure.

To avoid additional costs of such order a relatively simple method in which simply supported decks are tied together by locally debonded 'tying' bars at deck ends, has been used in some bridges during the mid eighties. Results of their performance are not fully know yet. However, complicated detailing and possible poor workmanship, the presence of a shrinkage crack at the throat and concentrated rotations at deck ends affecting surfacing, are potential causes for concern in the long term.

## 3 Proposed approach

The locally separated deck slab continuity concept as presented here avoids complicated design and construction and may even result in some savings below conventional jointed construction due to simplifications in detailing and drainage arrangements at these locations.

In view of the durability problems associated with the conventional jointed construction and great complexity of design and construction of 'live load' continuity decks, the author invented this concept about 10 years ago. The method is briefly described in references 1 and 2 and has now been implemented in some eight bridges in the UK including the 5 span 110m long river Frome bridge, with Department of Transport approval. It is now know that several other countries notably Germany have used similar ideas on large numbers of bridges apparently with considerable success. It is however a sad commentary on European Community co-operation that important technical matters of day to day concern to engineers are not being communicated across linguistic boundaries.

## 4 The conceptual basis

In typical single span precast beam bridge construction, a relatively thin top slab is customarily used to affect load distribution between the beams. Depending on load positioning, the beam elements rotate and deflect differentially to each other causing the deck slab to flex to accommodate such displacements. The structural action of the slab is complex and involves interactions of flexure, shear, torsion and some local effects. This load distributive action of the slab can be viewed as two simple modes of flexural actions viz (a) symmetric and (b) asymmetric bending of a pin ended slab element as depicted in Fig. 1. It will be noted that mode (b) here allows the relative deflections of the ends in relation to the deflected end tangents. Various deflected forms of a slab element due to end deflections or rotations are linear combinations of these two modes. In addition there is also the equilibrium action of the slab when loads are applied within its span.

The ends of bridge decks simply supported on metal or elastomeric bearings at intermediate supports undergo small rotations and deflection relative to each other depending on load positioning on the deck. Although the main spans of the bridge are large in comparison to slab spans between beams, the depth of construction and stiffnesses are large too,

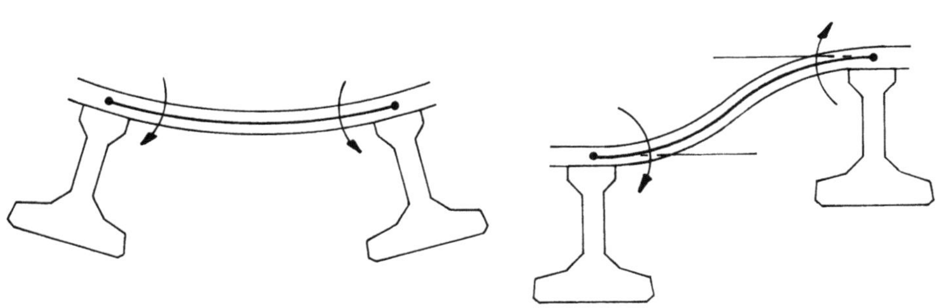

(a) Symmetric mode (b) Asymmetric mode

Fig. 1 Modes of flexure of transverse slab

resulting in similar order of rotations as for the slab. It therefore follows that slab panels of similar proportions as are conventionally used between the precast beams, should be able to accommodate movements at span ends, a matter to be confirmed by the detailed design process. It will be noted that this deck slab continuity by itself, will not cause any net forces in the connecting slab panels. Fig. 2 indicates the flexing of such a slab in the basic and combined modes.

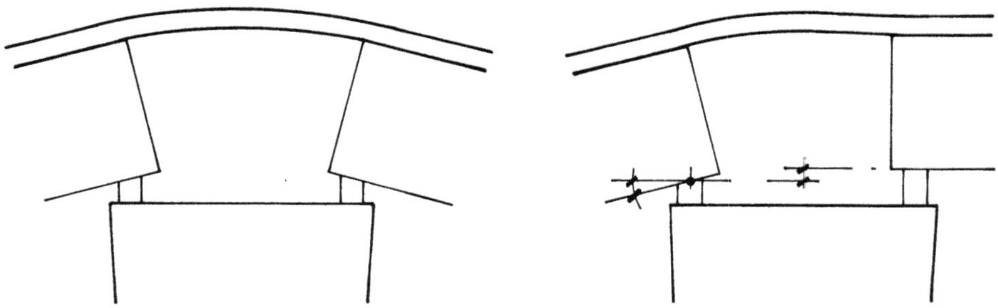

(a) Symmetric rotation (b) Relative deflection and rotation

Fig. 2 Mode of flexure of connecting slab panel at support

Under design loadings the main spans will deflect in the sagging mode and the connecting slabs in hogging mode. In the absence of live loads but due to temperature difference and creep effects the span elements could be hogging upwards and the connecting slabs sagging downwards. These two extreme conditions are indicated in Fig. 3. Connecting slabs will thus be flexing in reverse curvature to the main span members of the bridge. It will be noted that decks with such continuity of displacements between spans will provide a suitable base for laying waterproofing and surfacing for vehicular use.

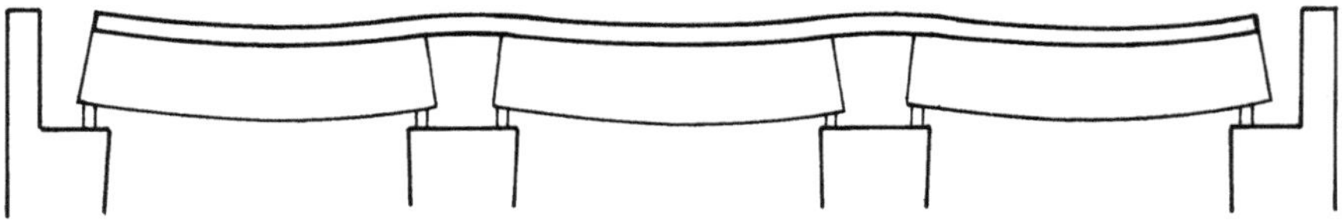

(a) spans sagging

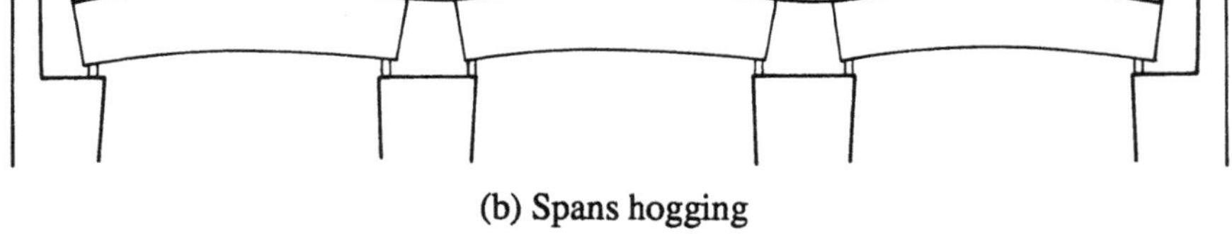

(b) Spans hogging

Fig. 3 Longitudinal action of connecting slab panels

## 5 Application

For practical applications of this concept to composite bridges the following aspects need to be considered:

(a) In order to accommodate connecting slab panels between spans without widening the supports, it would in general be necessary to locally separate the deck slab from the tops of beams, thus creating independently acting slab elements as shown in Fig. 4. The width of supports required for such arrangement would be the same as for simply supported construction.

(b) Transverse diaphragms should be provided set at connecting slab edges as shown in Fig. 4. This will minimise punching shears, allow easy jacking for probable bearing replacement, reduce relative displacements of adjoining beams and contribute to load distribution properties of the deck.

(c) Connecting slab may also be used at one of the ballast walls as shown in Fig. 5 and the entire longitudinal movement of the deck accommodated at the other abutment. For this option connecting slab should preferably be located at downstream end to allow shedding of water from the bridge deck without percolation through the joint. However locating connecting slab here, would reduce ability for jacking for bearing replacement therefore high stiffness bearings such as metal bearings requiring only small amount of lifting would be more appropriate here. For obvious reasons, connecting slabs at both abutments are not recommended for multi-span bridges.

(d) At significant parapet upstands at intermediate support positions, the upstands should be jointed and separated from the connecting slab as shown in Fig. 6. If however

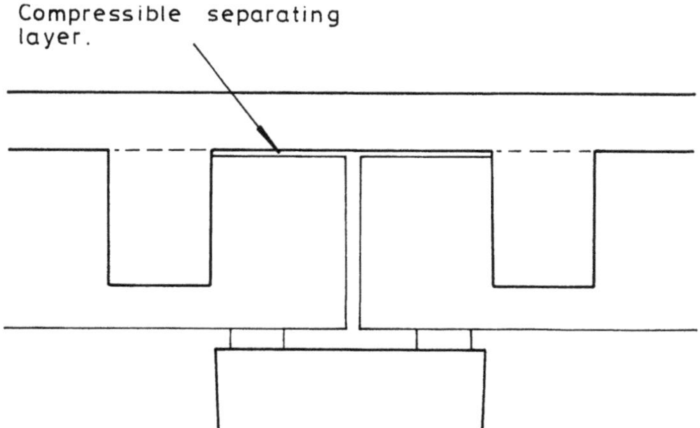


Fig. 4 Locally separated deck slab with transverse diaphragm

Fig. 5 Connecting slab also at one of the abutments

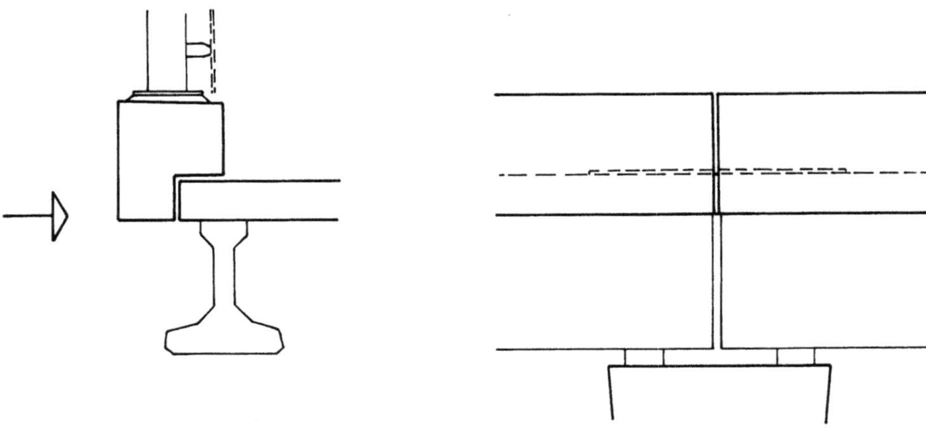

Fig. 6 Separating and jointing of parapet upstand

these upstands are not too deep or significant cantilevers or wide footpaths exist on the deck, it may be possible to control any cracking without the need for separating and jointing the upstands.

## 6 Design considerations

As explained before, the mode of flexural action of the connecting slabs is opposite to the main spans. Despite its ability to support local wheel loads, its contribution towards restraining the main spans against end rotations is relatively modest. It would therefore be simple and conservative to design the main spans without taking advantage of this restraint. The design of the connecting slabs can then be conservatively based on the calculated end rotations of the span members. The connecting slab proportions are generally such that these rotations will indicate cracking which should comply with Code requirements.

It therefore follows that should restraint advantage from the connecting slab be desired to be considered in the global analysis of the main deck, only the cracked stiffness of this component is used in the model, irrespective of cracked, uncracked or gross inertias being chosen for the representation of deck members in accordance with the Code.

The length, thickness and reinforcing of connecting slabs will normally depend on structural configuration, range of ages of beams, level of prestress, live loads, local wheel loads, stiffness of bearings, temperature differences, differential settlements, skew and method of articulating the bridge. This approach is equally applicable with the M, Y and U beam bridge decks currently used in the UK. Moderate lengths of connecting slab are likely to prove adequate as local and global effects will tend to counteract each other. Fatigue in reinforcement due to stress reversals, should be designed 'out' by compliance with Code requirements, as for the rest of the deck slab.

Preliminary calculations indicate that possibly up to 12 spans could be made continuous using this approach. However, it should be noted that as the number of spans increases, the forces due to friction or shear in the bearings and due to dynamic effects due to pulling of many spans, would tend to increase in the connecting slabs. Without detailed calculations for these effects, not more than about 6 spans should be made continuous using this technique.

## 7 European experience

As mentioned earlier it is now known that similar ideas have been applied elsewhere in Europe notably in Bavaria in Germany and to a lesser extent in Norway. In German practice outlined in reference 3 and Norwegian practice outlined in reference 4, this approach of slab connection is referred to as 'continuity plate' and 'isolation plate' construction respectively. Although the success largely depends on the detailed arrangements adopted, the experience of these countries with this approach is thought to be very satisfactory.

## 8 Advantages

The principal advantage of this approach is its great simplicity of design and construction combined with effectiveness. By thoughtful design this approach could result in first costs below simply supported construction due to simplification of details of diaphragms, joints, parapets and drainage system.

In the 'live load' continuity approach the cracking of the top slab adjoining the continuity diaphragm member is likely to extend to the full depth of the deck slab due to global tension arising here under live loads. In the proposed approach here, the connecting slab predominately acts in flexure therefore any cracking would tend to be limited to one face of this slab, the other face being in compression thus reducing the possibility of water penetration.

Because of flexible connection between spans, the maximum reactions likely to arise on the support elements could be less than 'live load' continuity approach. Also this approach would provide greater tolerance towards differential settlements of supports.

## 9 Conclusions

The approach described is of wide applicability in composite concrete bridge construction and has now been designed in 8 bridges in the UK including one with 30 degrees skew. For bridges to be designed as simply supported in compliance with the Codes, it may be possible to develop some standardised recommendations for the design of this element.

For steel/concrete composite bridge construction where steel beam continuity is not incorporated by splicing, the concepts described in this paper can be used with span by span construction to achieve deck slab continuity to advantage.

## 10 References

Kumar A., Continuity in composite concrete bridge construction. International conference on short and medium span bridges, Montreal, Canada, August 1986.

Kumar A., Composite concrete bridge superstructures. Wexham Springs, British Cement Association, 1988. 46pp. Publications 46.505.

Wagner F J., Nuremberg and Buchting F, Munich. Die Federplattenbauweise bei mehrfeldrigen Fertigtragerbrucken aus Beton (Continuity plate construction in multi-span bridges using pre-cast concrete beams), Bauingenieur, Springer Verlag 1981.

Normaler Handbok - 100, Bruprosjektering - 08 (Nib-Bruer) and - 09 (Nob-Bruer) '83, Statens Vegvesen, Vegdirektoratet, Oslo.

# REPORT ON SESSION SEVEN

G.P. TILLY

Mr Pritchard described his review of the techniques used to make precast prestressed beams continuous. He favoured the use of wide in situ cross heads because they can accommodate deck plan and vertical curvature and avoid the use of half-joints. The effects of creep and shrinkage balance out with time. Mr Pritchard cited the case of a bridge under construction in Kuwait when the beams being placed on the pier head caused it to deflect so that collapse occurred. A questioner commented that this system of construction was considered for Killikrankie Bridge; the stability of the pier top was investigated and a possible solution was to use an overhead gantry. In the event this procedure was not adopted. It was pointed out that it is necessary to consider the development of cracking in the soffit concrete over the support. Hybrid beams, combining straight pretensioned tendons and ducted post-tensioned tendons can be used as an alternative to draped pretensioned tendons.

Dr Kumar presented his paper and explained that with continuous connections the resulting sagging movements could be the same as for simply supported design but hogging movements are higher. It is possible to eliminate the end joint but the abutment has to slide or rotate to accommodate movement. Overall it is necessary to weigh up the complexity of the design against the advantage of eliminating expansion joints. He then explained the advantages of separating the deck slab over the supports and making that alone continuous. A speaker commented that in Germany it had been common practice to have slab only continuity between precast beams since the 1970s. It is possible to do this more economically than for simply supported construction because cheaper beams can be used.

The author responded that it is not surprising that the technique is used in Germany, the only surprise is that it has not been used more in the UK. It is feasible to use shorter beams but there are difficulties in reducing the sagging movements. There are also possibilities for using different designs of anchorage and shallower beams. It is necessary for the deck slab to be able to deform and follow the movements of the structural elements.

# PART EIGHT

# SOIL–STRUCTURE INTERACTION

# 22 SOIL–STRUCTURE INTERACTION: CENTRIFUGE MODELLING OF INTEGRAL BRIDGE ABUTMENTS

S.M. SPRINGMAN and A.R.M. NORRISH
Department of Engineering, University of Cambridge, UK

Abstract
Joint-free 'integral' bridges are now being recommended by cost-conscious transport authorities from around the world. Avoiding the problems associated with corrosion and mechanical disruption of expansion joints, by excluding them, creates a situation in which thermal strains induced in the deck cause cyclic loading on the soil behind the abutments. There are two aspects of soil behaviour which are of concern to the design engineer - whether the soil behind the wall will settle significantly and whether there is a build up in horizontal stresses which will cause a severe increase in the bending moment carried by the abutment wall.
Keywords: Soil-structure interaction, Integral bridges, Abutments, Temperature effects, Strain cycling, Non-linearity, Hysteresis.

## 1 Introduction

The Transport Research Laboratory have commissioned a series of centrifuge model tests to examine the behaviour of full-height 'rigid frame' or 'integral' bridges retaining cohesionless soil. The rationale for bridge design using this method is already well documented (Hambly, 1991; Pritchard, 1992).

Two different types of abutment will be investigated: a relatively flexible piled wall with a stiff deck (Figure 1a), or a stiffer spread base abutment combined with the same deck (Figure 1b). In the first case, the mode of behaviour is anticipated to be flexure of the pile wall combined with a rigid connection which forces rotation at the wall-deck joint. In the second example, the displacement at deck level will be accommodated by a combination of rotation, translational shear and perhaps a degree of flexure.

This paper will focus on some mechanisms which will affect the soil-structure interaction, the test specifications and a brief description of the equipment, before quoting some preliminary results.

*Continuous and Integral Bridges*. Edited by B.P. Pritchard. Published in 1994
by Taylor & Francis, 2 Park Square, Milton Park, Abingdon, Oxon, OX14 4RN. ISBN: 0 419 19030 9.

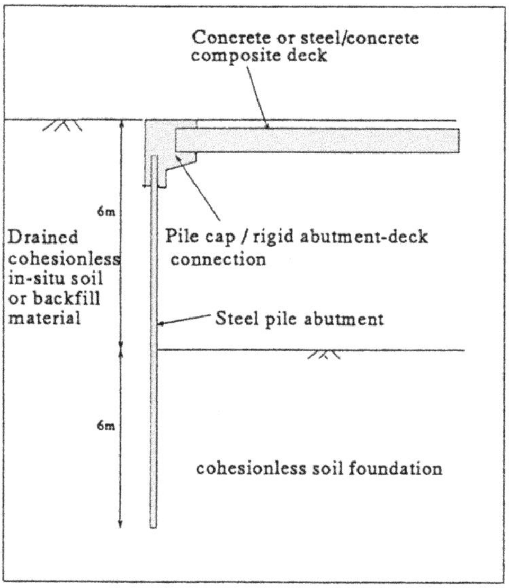


Fig. 1a. Piled integral abutment

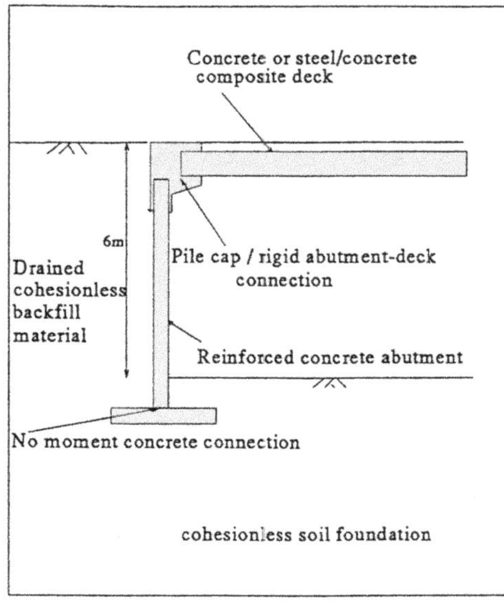


Fig. 1b. Spread base integral abutment

## 2 Centrifuge model testing: Bridge abutments

Table 1. Summary of centrifuge model tests on soil-structure interaction of bridges

| Date | Research Focus | Sponsor | Researcher |
|---|---|---|---|
| 1984-1987 (10 tests) | Single row of piles/pile groups in soft clay overlying sand adjacent to surcharge loads | TRRL | Springman (1989) |
| 1985-1988 (7 tests) | Spread base abutments on stiff clay, sand embankment poured in-flight | - | Sun (1989) |
| 1992-1993 (4 tests) | Buttonhole foundations - pile group adjacent to surcharge load | EC | Bransby (1992, 1993) |
| 1992- | Rigid frame bridges in sand | TRL | Norrish (1993) |
| 1992- | Piled full-height abutments, in soft clay overlying sand, sand embankment poured in-flight | TRL | Ellis (1993) |
| Reports: | | | |
| | Contractor Report 196: The effect of surcharge loading adjacent to piles | TRRL | Springman & Bolton, 1990 |
| | Contractor Report 270: Geotechnical stress analysis for bridge abutment design | TRRL | Bolton, 1991 |

The Cambridge Geotechnical 10 m beam centrifuge has been used to investigate the behaviour of several types of bridge abutments and piers (Springman, 1989; Sun, 1989) leading to the delivery of design recommendations (Springman & Bolton, 1990; Bolton, 1991). Bransby (1992), Ellis (1993) and Norrish (1993) are currently engaged in further research on the behaviour of full-height bridge abutments.

The advantages of modelling the soil-structure interaction at the correct stress-strain levels, and at small scale in the centrifuge, are many. The ability to observe the *deformation* mechanisms at *serviceability* limit state, and then the *failure* modes at *ultimate* limit state, is particularly valuable. The time scales and financial outlay are also significantly less than might be expected for an equivalent range of prototype full scale field tests. Schofield (1980) describes the principles of centrifuge modelling.

## 3 Behaviour of integral bridges

Diurnal and seasonal temperature variations will cause expansion and contraction of a continuous integral bridge deck, which, in the absence of an expansion joint, will be accommodated by cyclic lateral displacements of the wall-deck joint. As the abutment moves away from and towards the compliant fill, the cohesionless soil particles will tend to redistribute and repack themselves. If the wall movement away from the retained fill is large enough, the particles will be unable to maintain tension behind the wall and will tend to fall into any potential gap.

The soil-structure interaction mechanism is entirely strain-driven. Previous analyses seem to have opted for calculations based on elastic parameters, which were chosen to suit the operational range of stresses and strains (e.g. using a tangent shear modulus, $G_{tan}$, which is very different from the secant shear modulus $G_{sec}$, Figure 2c). However, the past stress history, the recent strain path and strain increment following the last reversal have a major influence on the response.

Sands will follow highly non-linear stress-strain behaviour (Figure 2a), and may show dilatancy (strain softening, Figure 2a; volume increase, Figure 2b), and further exhibit significant hysteresis and cyclic mobility (Figures 2c and 2d) during these cyclic deformations. A hyperbolic, hysteretic constitutive model is probably the most suitable.

Preliminary data on the soil response created by cycling between stress levels (Figure 2c) and strain or displacement conditions (Figure 2d) have been presented respectively by Bassett and Powrie (1992) and Bassett et al. (1992). The latter case is relevant to this class of problem and will be referred to in greater detail during the current research programme.

The resulting stress path followed by an element of soil behind the wall due to the strain cycling is highly complex - a combination of compression, extension and shear, and will be investigated further in due course. The element will be expected to undergo some rotation in principal stresses, and this will also be affected by the degree of adhesion $\delta$ obtained on the soil-wall boundary. In the field, this soil-wall boundary would be quite rough, and $\delta_{mob}$ might be close to the critical state value for the soil, $\phi'_{crit}$. In the event of a wedge failure behind the wall, the inclination of the failure plane will either depend on the value of $\phi'$ if the condition of maximum stress obliquity rules, or the lines of no extension, in which case dilatancy $\psi$ is important.

The horizontal earth pressure coefficient is likely to build up, maybe even approaching passive levels in really extreme cases. As the abutment is cycled into and away from the fill, this 'ratchetting' behaviour will amplify the bending moments in the wall, and significant settlement in the region approaching the bridge deck. Both of these mechanisms have implications for the serviceability of the bridge. Cracks may form in the abutment wall and the road surface will settle.

Cohesionless soils are most likely to be used in practice to construct the approach embankment, and therefore have been specified for this study. There is a greater probability of damage to the abutment given the combination of particular mobility and the stiff response of compacted granular fills. Furthermore, the peak angle of shearing resistance will be higher than typical values for clays, leading to a potentially larger lateral pressure coefficient and more horizontal stress on the abutment wall.

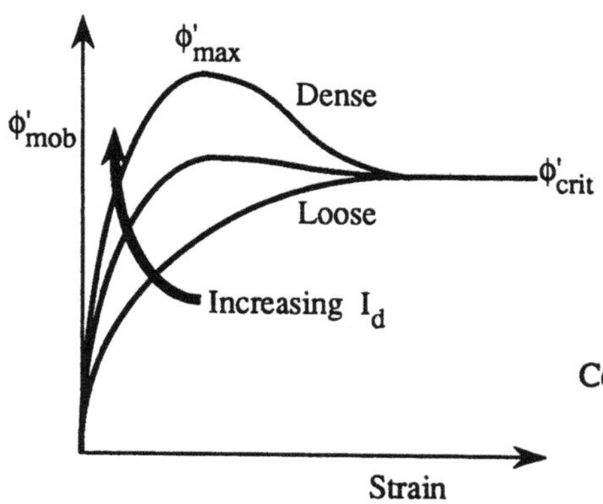


Fig. 2a. Influence of $I_d$ on $\tau$–$\gamma$ response

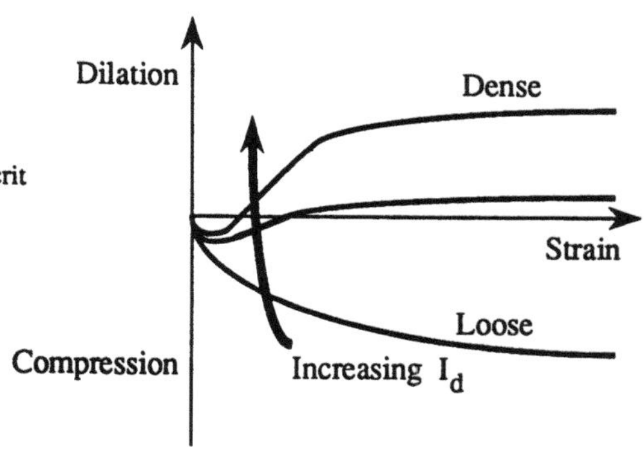


Fig. 2b. Influence of $I_d$ on volume change

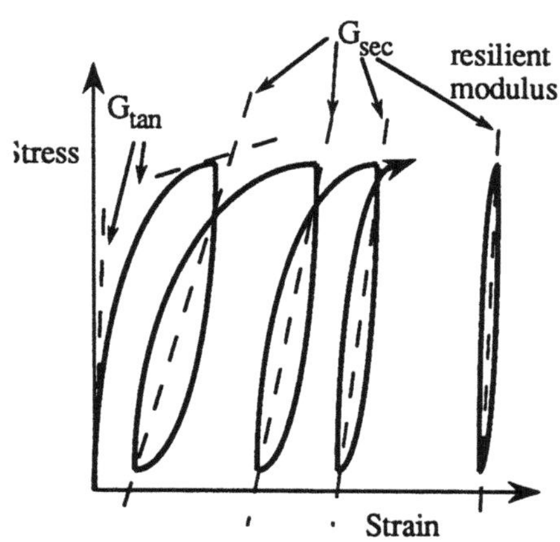


Fig. 2c. Constant stress cycling

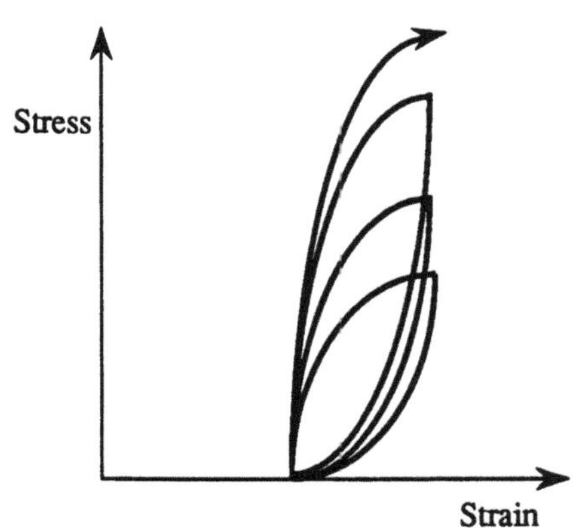


Fig. 2d. Constant strain cycling

The influence of number of cycles N of deformation on the crucial parameters will be of interest to the designer. Do the wall bending moments and pressures approach a constant value? Is the settlement behind the wall limited after a specific number of cycles? Is there any variation in the behaviour if the abutment was initially perturbed in an active (away from the retained fill) or a passive (towards the retained fill) sense?

Table 2: Prototype specification and scaling factors

| Variable | Field prototype | Scale | Centrifuge model |
|---|---|---|---|
| Bridge span | 50-150 m | 1/60 | 200 mm |
| Cyclic movement | +/- 5 mm diurnally, +/- 30 mm annually | 1/60 | +/- 0.1 mm diurnally, +/- 0.5 mm annually (also test +/- 1 mm) |
| Retained height of fill | 6 m | 1/60 | 100 mm |
| Embedment for piled wall | 6 m | 1/60 | 100 mm |
| Piled wall flexural stiffness | 1.5 $10^5$ $kNm^2/m$ | $1/60^3$ | 0.69 $kNm^2/m$ |
| Spread base wall flexural stiffness | 1.5 $10^6$ $kNm^2/m$ | $1/60^3$ | 6.9 $kNm^2/m$ |
| Deck flexural stiffness | 4.5 $10^6$ $kNm^2/m$ | $1/60^3$ | 20.8 $kNm^2/m$ |
| Sand particle size | 5.4-9 mm | 1/60 | 0.09-0.15 mm |
| Piled wall thickness | | 1/60 | 4.92 mm |
| Depth of deck | | 1/60 | 18.2 mm |
| Bending moment / m width | | $1/60^2$ | |
| Shear force / m width | | 1/60 | |
| Lateral pressure | | 1 | |

## 4 Structural design of the centrifuge model

The dimensions of the field prototype, together with the centrifuge model equivalent are given in Table 2. Since these tests are to be carried out at a nominal acceleration of 60 gravities, then there is a linear scaling factor of 60 on all model dimensions. The factors for flexural stiffness, bending moment and shear force per metre length of wall are also quoted in Table 2.

The general arrangement of the centrifuge model is shown in Figure 3. The wall was made out of 6061/T6 grade Dural and the deck out of mild steel. There was a rigid moment-carrying connection at the wall-deck joint. The displacement cycles were carried out using a stepper motor driving an eccentric rotating cam system through a vertical pin joint. This joint was also supported against vertical movement downwards (but not upwards) at the end of the deck.

A full 360° rotation of the cam, without allowing for the small degree of backlash, gives peak to peak displacements of +/-1 mm at model scale (1/60), varying sinusoidally, which becomes +/- 60 mm at prototype scale. The cycles appear to be uniform and repeatable at each set of nominal displacements. For cam rotations less than 360° , e.g. for +/- 0.1 mm (+/- 6 mm prototype) and +/- 0.5 mm (+/- 30 mm prototype), the 'waveform' was sawtooth in shape.

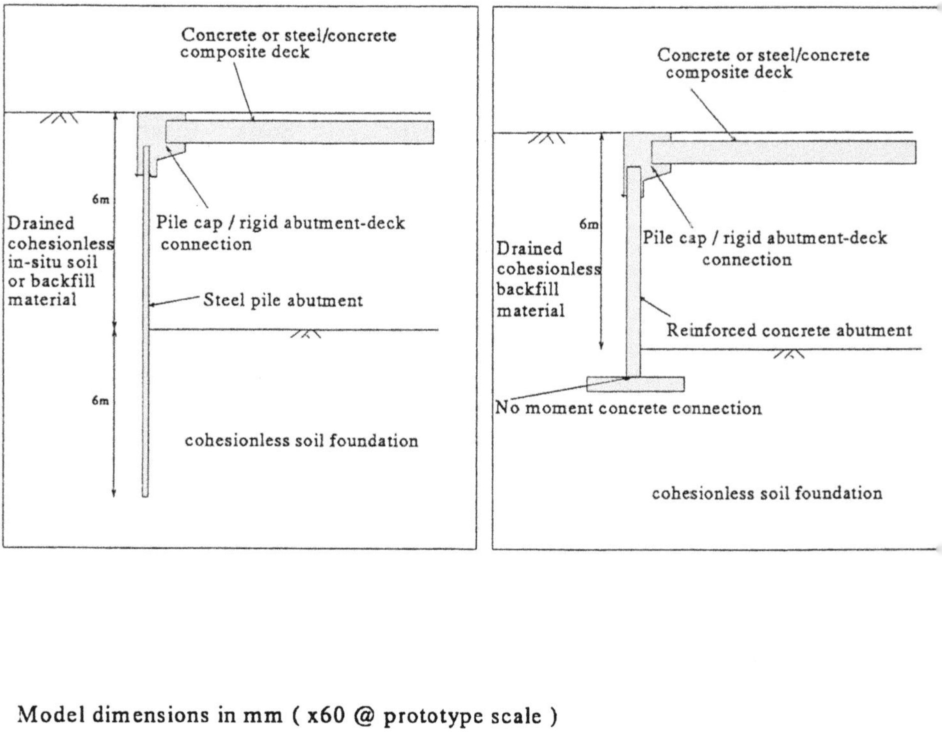


Fig. 3. Layout of the centrifuge model

## 5 Modelling the soil

The choice of grain size for the model soil is critical when the mechanisms that will obtain at particulate level are to be replicated. The smallest grain size that it is practical to work with is silica sand, 100/170 grade (fraction E), with a particle size ranging between 90 - 150 μm (5.4 - 9 mm at prototype scale). Therefore, for the initial set of strain cycles at +/- 6 mm, extensive repacking and vertical settlement is less likely to occur behind the wall.

The centrifuge models will be made from dry sand since the strain cycling is so slow that pore pressure response is unlikely to be significant - in contrast to the build up associated with dynamic events such as earthquakes. It is thought to be important to isolate the soil behaviour at particulate level at this stage, given the complications that will arise with maintaining an elevated water table behind the wall. Should significant leakage occur, the balance of the centrifuge is in jeopardy.

The model was prepared by raining sand into the centrifuge strongbox in order to achieve a relative density $I_d$ of about 80%. The $I_d$ of this fill will have a major influence on the soil-structure interaction. A zone of loose fill behind the abutment will be less likely to cause lateral pressures to build up towards passive levels. The Americans favour this method, and use an approach slab to span the zone that has settled in behind the abutment. Other authorities prefer to specify a denser, compacted fill.

The effect of different densities of the fill behind the wall will be one of the variables examined in this series of tests. Relative density is also an important indicator of the potential dilatancy $\psi$ of the soil, which will elevate the mobilised angle of friction, $\phi'_{mob}$ well above the critical state value $\phi'_{crit}$ (usually about 33° for quartz), so that in the extreme case $\phi'_{mob}$ becomes $\phi'_{max}$.

$$\phi'_{max} = \phi'_{crit} + A\,\psi \qquad (1)$$

A is a constant which is often taken as unity, although Bolton (1986) recommends a value of 0.8 taken from back analysis of series of triaxial and plane strain test data. $\psi$ may be determined, in degrees, from the following formulae (Bolton, 1986) and it can be seen that the dilatancy is increased at low mean effective stress p' (in kPa) and when $I_d$ is high:

$$\psi = 3\,\{\,I_d\,(10 - \ln p') - 1\,\} \qquad \text{(triaxial)} \qquad (2a)$$

$$\psi = 5\,\{\,I_d\,(10 - \ln p') - 1\,\} \qquad \text{(plane strain)} \qquad (2b)$$

A higher peak value of passive earth pressure, with coefficient $K_p$ is then implied (at the point when the major principal stress direction has rotated to the horizontal):

$$K_p = (1 + \sin\phi'_{mob})\,/\,(1 - \sin\phi'_{mob}) \qquad (3)$$

increasing as $\phi'_{mob}$ approaches $\phi'_{max}$ before dropping off to $\phi'_{crit}$ (Figure 2a), with the associated variation in lateral earth pressures and wall bending moments.

## 6 Test procedures

The wall was held in position while the sand was poured at 1 g, and so the stress path did not entirely resemble the full scale equivalent. Craig (1984) noted that installation of model piles at the correct stress levels was far more critical for investigation of axial response than for lateral behaviour, which is the mode of perturbation applied here. A typical prototype horizontal stress profile acting on the abutment wall following compaction of the fill during construction is given in Figure 4.

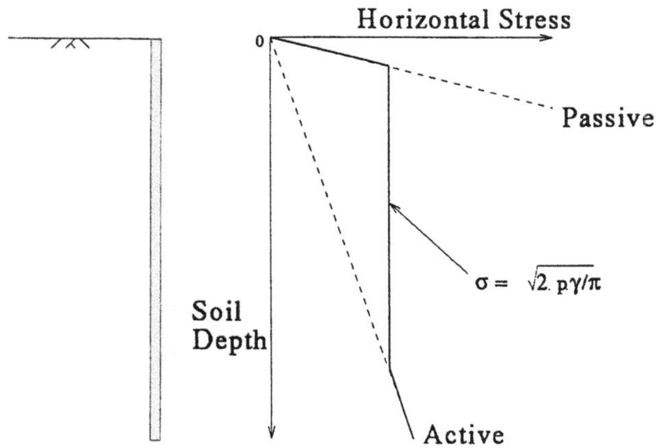


Fig. 4. Distribution of horizontal stresses in compacted backfill in the field

Following assembly of the actuator and connection of the displacement transducers and bending moment strain gauge bridges, the model was loaded onto the centrifuge and all systems were checked again from the centrifuge control room. The package was then subjected to a brief load cycle created by accelerating the centrifuge model to 100 g so that the soil was overstressed by a factor of 1.67 relative to the values expected at a nominal acceleration of 60 g, and the earth pressure coefficient was artificially increased.

Three sets of displacement cycles were carried out:

(a) N = 50 cycles at +/- 6 mm, prototype scale
(b) N = 75 cycles at +/- 30 mm, prototype scale
(c) N = 100 cycles at +/- 60 mm, prototype scale (not reported here).

In each case, an initial displacement of the same amount as the subsequent displacement cycles was carried out in either an active sense (away from the fill) or in a passive sense (towards the fill). Data was taken from bending moment transducers in the walls and deck, together with the axial load cell in the deck and various displacement transducers.

In this paper, a linear approximation of the bending moment diagram will be integrated twice to give a profile of displacement down the wall, which may be fitted to boundary conditions obtained from the displacement transducers to check internal consistency of the data. Similarly, the bending moment profile may be differentiated once to give an approximate shear force distribution for comparison with the axial load reading at deck level. In the foreseeable future, the data will be fitted by either a polynomial or a spline, whereupon approximate pressure distributions will be obtained by differentiating twice.

## 7 Preliminary results

Preliminary analysis of data from two centrifuge model tests conducted last week will be presented below. All data is presented at prototype scales.

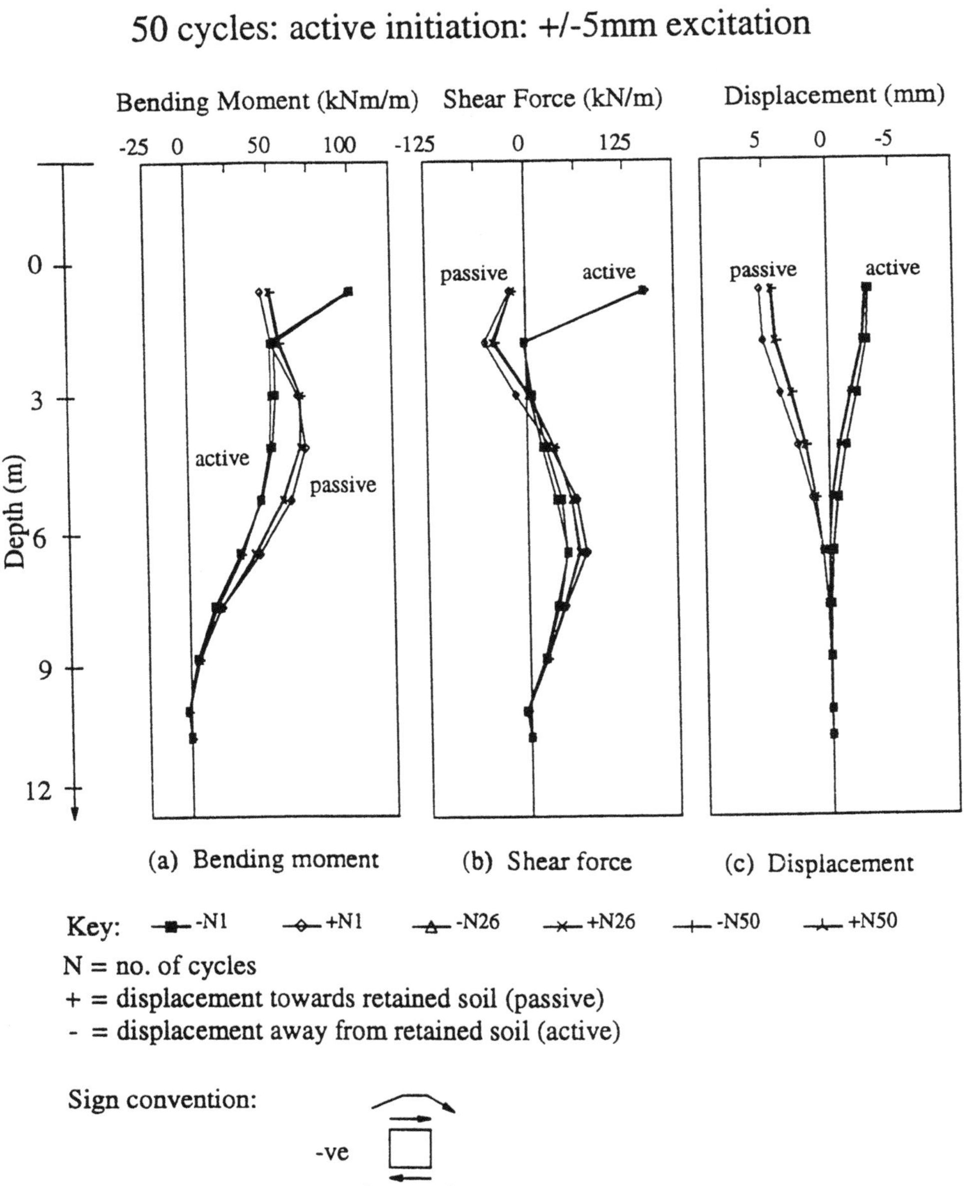


Fig. 5. Data from the active initiation of a smooth wall over 50 cycles of +/- 6 mm

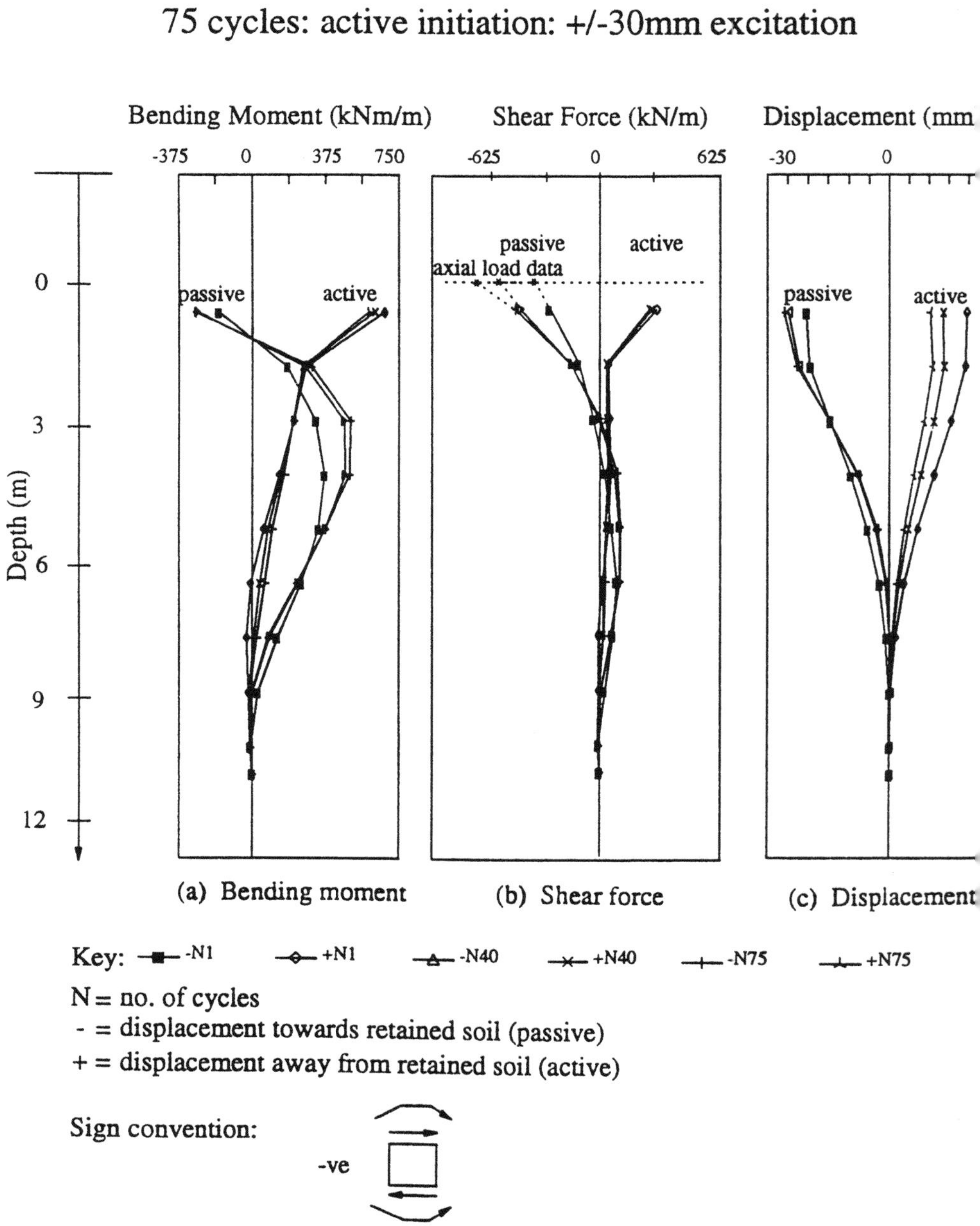


Fig. 6. Data from the active initiation of a smooth wall over 75 cycles of +/- 30 mm

### 7.1 Small displacement behaviour

The bending moment and shear force per unit width, together with wall displacement are given in Figures 5a, 5b and 5c, for 50 cycles of +/- 6 mm excitation at the deck. There appeared to be minimal change in the bending moments and shear forces with increasing numbers of displacement cycles, and also not much difference between data from both the active and the passive modes of deformation. The point at which the bending moments begin to amplify with increasing numbers of cycles requires further investigation, but it would seem that the multiple diurnal cycles of expansion and contraction are unlikely to cause significant distress to the wall. The maximum vertical settlement in the fill behind the wall was less than 12 mm for a zone 0.4 m long.

### 7.2 Large displacement behaviour

75 cycles of +/- 30 mm displacement created a wedge formation in the sand behind the abutment wall. For the test in which excitation was initiated by an active displacement, the wedge formed over almost the full height of the retained fill at an angle of approximately 15° to the vertical, and this was associated with significant settlement of 210 mm at sub-pavement level.

The number of cycles did not significantly affect the bending moments (Figure 6a) generated by *active* displacement of the wall (indeed the peak moment occurring at the 1st cycle was the greatest). However, the peak bending moments due to successive *passive* cycles increased quite noticeably and migrated to a shallower depth, implying that the centroid of lateral pressure was also moving upwards. The cumulative and incremental escalation in bending moment together with an average % increase per cycle over 75 cycles is summarised in Table 3. After 40 cycles, the rate of accumulation is significantly reduced, but not eliminated altogether.

The shear force diagram (Figure 6b) demonstrates the internal consistency of the data. A linear fit to the bending moment profile has been differentiated to give a shear force distribution for 1, 40 and 75 cycles of active and passive excitation. These values are compared with the axial load measured in the bridge deck by a series of transducers. Agreement seems to be excellent.

Table 3. Increase in bending moment for N cycles in the passive mode; active initiation

| Cycle No. | % increase in bending moment | | |
|---|---|---|---|
| (N) | cumulative | increment | average % / cycle |
| 2 | 9.0 | 9.0 | 9.0 |
| 5 | 14.4 | 5.4 | 1.8 |
| 10 | 18.5 | 4.1 | 0.8 |
| 20 | 23.9 | 5.4 | 0.5 |
| 40 | 29.9 | 6.0 | 0.3 |
| 75 | 31.6 | 1.7 | 0.05 |

The displacement profile of the wall (Figure 6c) confirms the double integration of the linear fit to the bending moments (values measured directly from the displacement transducers are omitted for clarity). It is important to note that there is neither rigid body rotation, translation, shear force nor bending moment at the tip of the abutment wall.

An initial scan of the data from the second centrifuge model test in which the initial excitation was in the passive direction indicates very similar bending moments in magnitude and shape, and comparable values of compressive load in the deck. However, further detailed analysis will be required to explain why the wedge mechanism was smaller, both in depth and the volume of 'mobile' fill.

## 8 Future work

Aside from more detailed analysis of the data presented here, the variables to be investigated in these and future centrifuge model tests from this series will be:

(a) density of fill
(b) methods of compaction
(c) direction of initial perturbation
(d) soil-wall adhesion
(e) flexural stiffness of the wall.

Amendments are planned to the existing test package to give more direct readings of pressure on the wall, and some information about in-situ, small strain soil stiffness.

Numerical analyses will be conducted in parallel to these tests to calibrate the model data, with assistance from soil parameters obtained through plane strain stress path tests that have replicated the stress history experienced by soil elements in the centrifuge.

## 9 Conclusions

The soil-structure interaction for an integral bridge is highly complex and will not be well modelled by elasticity. Several aspects must be considered to explain cohesionless soil behaviour due to cyclic strains: particle repacking, dilatancy, hysteresis, cyclic mobility, and stress paths due to repetitive compression, shear and extension.

Centrifuge model tests offer a cost-effective opportunity of investigating these modes of behaviour by producing data of axial deck load, wall bending moments and displacements, as well as exhibiting the deformation and failure mechanisms. Key parameters may also be varied as required.

Large strain cycles create an increase in wall bending moment combined with a clear wedge shaped failure mechanism behind the wall, which is associated with severe slumping (0.21 m) of the backfill. After 75 cycles at +/- 30 mm horizontal displacement at the wall-deck joint, the peak values of bending moment had increased by nearly 32%, and the rate of accumulation had not completely levelled out. However, the change was minimal for the smaller cycles of +/- 6 mm which replicate the extreme diurnal variation.

## 10 Acknowledgements

The first author of this report is employed by Cambridge University and the second author is in receipt of a grant from the SERC. The work reported herein was carried out under a contract placed on them by the Transport Research Laboratory. The views expressed are not necessarily those of the Department of Transport.

The authors are also grateful to input from Dr Derek Carder at TRL, and to the design and technical staff at the Cambridge Geotechnical Centrifuge who helped to turn the ideas into reality, and to Dr Charles Ng for reviewing the manuscript and without whom the data would not have been ready for this paper.

## References

Bassett, R.H. and Powrie, W. (1992) The behaviour of granular materials. Course for industry. MACE Centre, Imperial College.

Bassett, R.H., Dalili, A. and England, G.L. (1992) Stress changes in granular materials subjected to controlled cyclic strains. Course for industry. MACE Centre, Imperial College.

Bolton, M.D. (1991) Geotechnical stress analysis for bridge abutment design. Transport & Road Research Laboratory Contractor Report, CR270.

Bolton, M.D. (1986) The strength and dilatancy of sands. **Geotechnique** 36, No. 1, 65-78.

Bransby, M.F. (1992) Piled foundations adjacent to surcharge loads. Cambridge University 1st year Report. 50p.

Bransby, M.F. (1993) Centrifuge test investigation of the buttonhole foundation technique. Data Report. 33p.

Craig, W.H. (1984) Installation studies for model piles, in **The application of centrifuge modelling to geotechnical design**. (ed. W.H. Craig), Engineering Department, Manchester, pp. 440-455.

Ellis, E.A. (1993) Lateral loading of bridge abutment piles due to soil movement. Cambridge University 1st year Report. 27p.

Hambly, E.C. (1991) **Bridge Deck Behaviour**. 2nd edition, E & FN Spon., UK.

Norrish, A.R.M.. (1993) Soil-structure interaction in rigid frame bridges. Cambridge University 1st year Report. 26p.

Pritchard, B. (1992) Advantages of bridge deck continuity, in **Bridge Design for Economy and Durability**. Thomas Telford Publications, London, UK.

Schofield, A.N. (1980) Cambridge University Geotechnical Centrifuge Operations. 20th Rankine lecture. **Geotechnique** 30, No. 3, 227-268.

Springman, S.M. (1989) Lateral loading on piles due to simulated embankment construction. Cambridge University PhD thesis.

Springman, S.M. and Bolton, M.D. (1990) The effect of surcharge loading adjacent to piles. Transport & Road Research Laboratory Contractor Report, CR196.

Sun, H.W. (1989) Ground deformation mechanisms for soil-structure interaction. Cambridge University PhD thesis.

# 23 THE CONTRIBUTION OF THE ROAD PAVEMENT IN ABSORBING HORIZONTAL MOVEMENTS IN AN EMBANKMENT

A.M. LOW
Ove Arup & Partners, London, UK

Abstract
This paper reports the results of a Finite Element computer analysis of a bridge abutment together with its approach embankment and the natural ground. The abutment is subjected to reversing horizontal displacements. Two runs were made, one with and one without the additional stiffness of the road pavement on the embankment. The differences in the results of the two runs are discussed.
Keywords: Abutment, Pavement, Finite element analysis, Temperature movements, Mohr-Coulomb

## 1 Introduction

Bridge designers face the same choice as the designers of railway tracks and highway pavements: Should the effect of temperature strains be controlled with movement joints at chosen centres or should the strains be resisted by elastic stresses, resulting in large end reactions? As far as I know those authorities which have built long integral bridges (say, longer than 35m) have provided end details with a transition slab spanning over uncompacted fill and a joint somewhere between the bridge and the road pavement. This analytical study considers the consequences of following the other route.

A study which was reported last year, Low (1992), is reworked and extended by comparing the behaviour of the abutment with and without a road pavement behind it.

## 2 The jointless bridge

We can start with some simple calculations: The temperature of a concrete bridge might rise by 27°C between its construction in the winter and its hottest day in service. If its length is fully restrained and its section is uniform then the stress in compression would rise about 10 MPa. This is high stress to absorb on top of the dead and live load stresses. It would be much easier to cater for the fully restrained stress of less than 5 MPa which results from the use of lightweight concrete.

*Continuous and Integral Bridges.* Edited by B.P. Pritchard. Published in 1994
by Taylor & Francis, 2 Park Square, Milton Park, Abingdon, Oxon, OX14 4RN. ISBN: 0 419 19030 9.

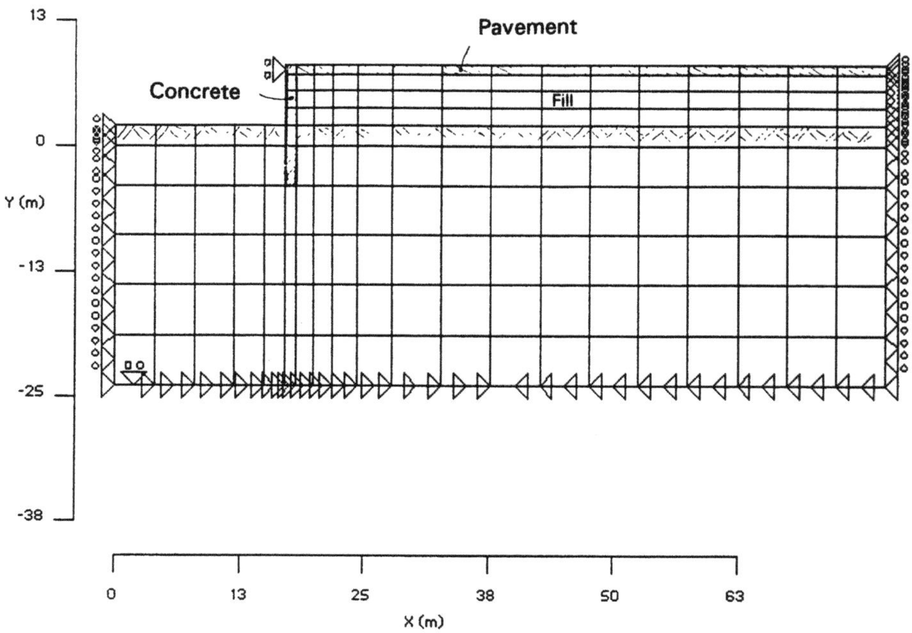


Fig.1. Finite element model

For a steel/concrete composite bridge the temperature rise might be 35°C giving a fully restrained stress of 86 MPa. This also is large if the steel starts in compression although, with care, a haunched beam shape can be chosen which greatly reduces those temperature compressions which combine with existing compressions.

In practice these values could only be relevant if the abutments were secured to rock foundations. Abutments with soil foundations provide partial restraint to the decks, they will move under temperature changes and the stresses will be much lower. It is the interaction between the deck forces, abutment movements and the stresses and strains in the soil that I have been studying with a finite element analysis.

## 3 Finite element analysis

Fig.1 shows a finite element half model of the soil around a symmetrical two-span overbridge for a dual three lane motorway. The overall length of the concrete structure is 36m and there is 60m of embankment behind the abutment. The embankment is 6.2m high of which the top 1m is described as "pavement". The abutment is a concrete wall 1m thick and 12.2m high, including its foundations. Bedrock is 26m below road level.

Lateral displacements are applied 0.5m below the top of the abutment wall. The model is analysed with SAFE (Soil Analysis using Finite Elements) which is a non-linear program from Oasys Ltd with

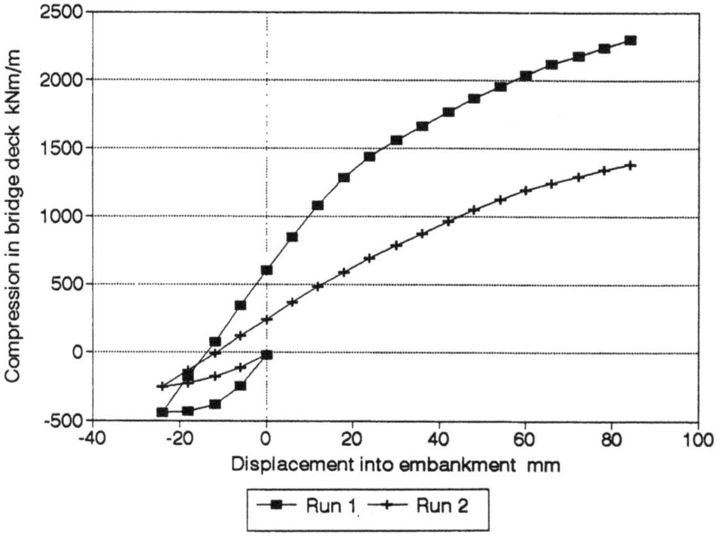


Fig.2. Force/displacement plot

several options for different soil models. It was run with elastic-plastic Mohr-Coulomb models for the ground, the embankment fill and the pavement. Initial compaction pressures were specified.

The principal material properties assumed are:

| | Zone | E (kPa) | ϕ' | c' (kPa) |
|---|---|---|---|---|
| Concrete | 2 | 30000000 | - | - |
| Pavement | 4 | 1000000 | 38 | 200 |
| Fill | 3 | 80000 | 36 | 0 |
| Natural ground | 1 | 40000 | 34 | 0 |

Two runs were made. The only difference between the two runs was that for Run 2 the material properties for the pavement zone were changed to those for the fill zone. Hence the model represented an earthwork with no road on top. Fig.2 shows the force/displacement plot for each run. The displacement was imposed in 6mm steps, a forward movement of four steps, away from the embankment, and then a backward movement of eighteen steps (108mm) into the embankment. The plots of each of the four movements indicate some degree of plastic softening after an initial elastic response. In both the forward movements the linear, elastic range appears to finish somewhere near the end of the first step and then the effect of yield is quite marked. In the backwards movements the deviation from linear behaviour is very gradual. In both runs strict linearity is lost after four or five steps but in the next two or three steps the deviation is slight and probably represents a condition which can be accepted infrequently by the embankment without impairing its performance.

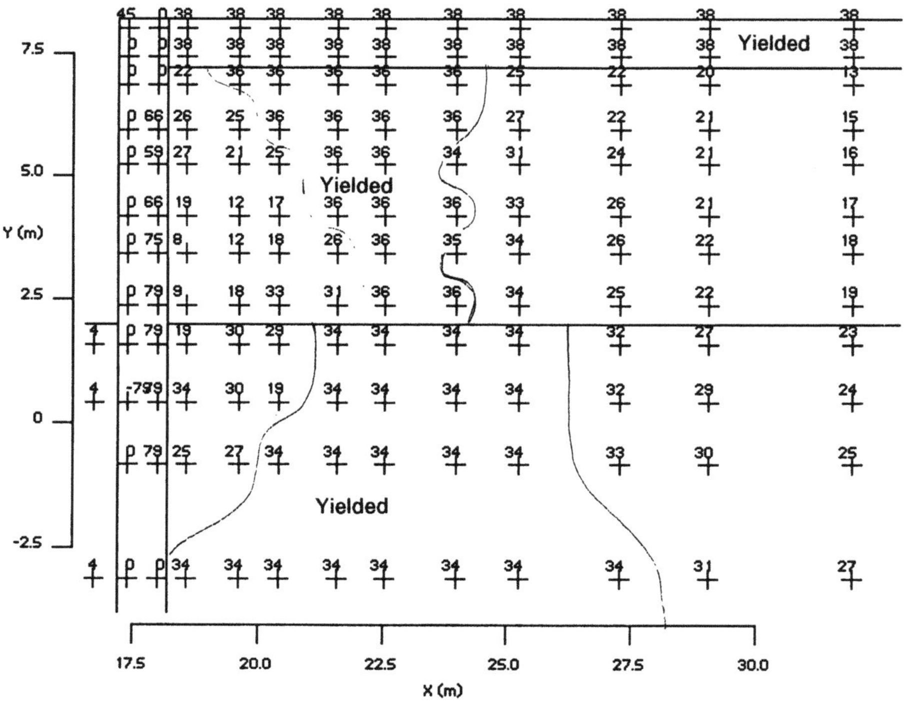


Fig.3 Extent of yielding at the end of Run 1

The purpose of the study is to derive some indication of the movement range which can be sustained in service so it is important to understand the physical situation which results in different degrees of deviation from the extrapolation of the linear behaviour. The SAFE program offers many formats for displaying the variation of stresses and other variables within the soil mass. The numeric display of "phi mobilised" gives a record of the extent of yielding at any stage (Fig.3). Phi-mobilised is the greatest friction angle (deviation of the force resultant from the normal) on any plane through a given point. Where yielding is occurring its value is the limiting phi of the soil so zones of yielding can be identified. The extent of the yielded zones is an indication of the degree of damage being suffered by the ground but it is an imperfect indication because it does not show the intensity (rate of straining) of the yield. This is only recorded by the program as the resulting cumulative displacement so the load/displacement plots are probably the best available indicator of damage.

Fig.4 Force/displacement interaction diagram

## 4 The results

The principle result can be seen in Fig.2. The extent of the linear response is similar for both runs and is about 36mm. It had been conjectured that the stiffness of the pavement would provide a force path to deliver some of the abutment thrust into the embankment well back from the bridge. For a given stress level the strains would be mobilised over a longer length resulting in larger displacements. The results do not support this conjecture, presumably because the additional stiffness increases the total force and counters the effect.

However the additional force is useful in the quest for long, joint-free bridges because it doubles the stiffness of the abutment and thus increases the proportion of the temperature movement that can be absorbed in deck strains.

The abutment stiffnesses from the linear range of the two runs are 43.1 kN/mm and 20.6 kN/mm. A concrete bridge deck might have a mean concrete sectional area of about 0.45 $m^2/m$ and be subjected to a temperature range of 47°C. Hence, with the pavement, the maximum deck length which could operate within the 36mm movement range would be 163m. Without the pavement it would be 142m. The equivalent figures for a deck in lightweight concrete would be 313m and 225m.

The diagram used to derive these values is shown in Fig.4. It demonstrates the interaction between the force in the deck and the displacement at each abutment.

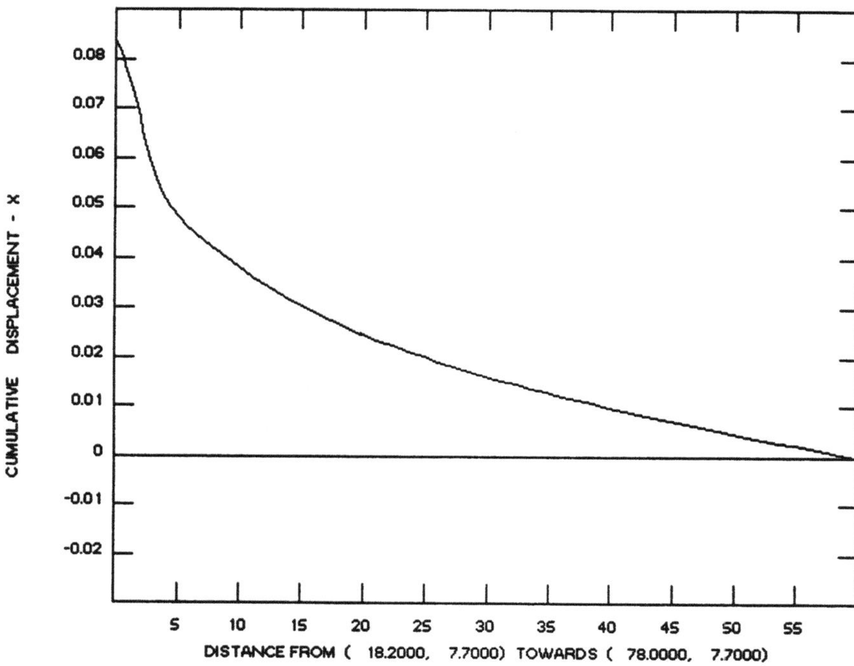


Fig.5 Horizontal displacements in the pavement at the end of Run 1

Results were also plotted for the variation of the horizontal displacements behind the abutment. Fig.5 shows the plot at maximum displacement for Run 1. The equivalent plot for Run 2 is similar. This shows that the strain (the slope of the plot) has not fallen to zero at the end of the model so the model size would affect the stiffness. In reality the stresses would dissipate into the third dimension and the non-linear behaviour would also reduce the effect of more distant strains. Hence a bigger model would probably not give a better stiffness.

## 5 Commentary

These results should not be applied to bridge designs without taking full account of the effects of the large restraint forces. It should be recognised that the restraint force will follow a straight line in plan between the elastic centres of the two abutments and care is needed in planning a skew bridge or one with a curved alignment. The digging of service trenches across the highway, close behind the abutment, would release some of the temperature restraint. Some control over such activities would be required.

## 6 Conclusions

The results of this study indicate that the inclusion of the horizontal stiffness of the highway pavement in the analysis allows joint-free bridges to be designed which are 15% longer than they would otherwise be. In general lightweight concrete joint-free bridge decks can be longer and in particular they derive a proportionately greater benefit from the additional restraint due to the road pavement. The increase is about 40%.

## 7 Reference

Low, A.M. (1992) Concepts in the design of the abutment in integral bridges, in Bridge Design for Durability, Expansion Joints and Continuity (ed Stephen Raggett), Transport Research Laboratory, Crowthorne.

# 24 THE INTERACTION BETWEEN BRIDGES AND EMBANKMENTS AND THE USE OF RUN-ON SLABS

S. THORBURN
Thorburn Ltd, Glasgow, UK

Synopsis
The interaction between earthwork embankments and bridges has been of interest to engineers over many years because of the incongruity of bridge structures on piled foundations and approach embankments resting on soft clay soils. The incompatibility of vertical displacements between bridge structures and adjacent embankments creates problems in respect of road pavement performance and maintenance.

Opportunity was taken during motorway construction in Scotland in 1976 to conduct extensive geotechnical investigations and to install instrumentation to monitor the performances of embankments in close proximity to bridge structures.

The paper examines the conclusions which may be drawn in respect of incompatibility of movements and the appropriateness of run–on slabs. The effects of lateral soil displacements on a bridge structure also are presented in the paper.
Keywords : Interaction, Embankments, Bridges, Clays, Geotechnical, Instrumentation, Run–on–slabs, Displacements.

## 1 Introduction

The two prime structures in the design of motorways and trunk roads are embankments and bridges.

These components are integral with each other in respect of performance and it is insufficient to design them as separate elements.
Concrete or steel bridges (hardware) and soil embankments (software) interact with each other and bridge designers cannot afford to ignore the interactive effects.

The subject of interaction is covered in the publication on the Real Behaviour of Structures published by the Institution of Structural Engineers in March 1989.

It is the experience of the author that there is an over–emphasis on the performances of embankments with lesser attention being given to the performances of bridges.

The major problem confronting the designer of embankments to be formed on very soft cohesive soils is not the mechanics of stability analyses,since these are well established, but rather the difficulty of estimating for the purposes of analysis the probable magnitudes and rates of dissipation of the excess porewater pressures generated by the construction of the embankments. The magnitudes and rates of dissipation must be predicted with reasonable accuracy

*Continuous and Integral Bridges.* Edited by B.P. Pritchard. Published in 1994
by Taylor & Francis, 2 Park Square, Milton Park, Abingdon, Oxon, OX14 4RN. ISBN: 0 419 19030 9.

to permit proper definition in contract documents of the sequence and rates of embankment construction. Contractors cannot efficiently plan their major earth–moving operations if programmes of construction cannot be formulated on the basis of reliable geotechnical predictions.

Pause clauses should be incorporated in tender documents and the word pause refers to the period of time between each stage of construction of an embankment during which earth–moving operations are discontinued to permit excess porewater pressures to dissipate to acceptable residual values. The extent to which judgement is used in pause clause predictions depends on the availability of characteristic data, and, if the volume of proper data is relatively small and variable then statistical theory cannot be applied and greater reliance must be placed on judgement.

A case history is presented in this paper for a high embankment founded on post–glacial deposits which comprised very soft carse clays.

Although the expenditure on the site investigation work for the motorway project was significant, the variations in the permeability and compressibility characteristics of the soft alluvium necessitated considerable judgement in the selection of design parameters. Field permeability tests were carried out to determine the coefficients of consolidation to be used for the pause clause predictions and extensive instrumentation was installed to permit the performances of the embankments and bridges to be monitored.

This paper examines the performance of an embankment and an adjacent bridge structure constructed some 16 years ago and discusses the relevance of the case history to the design of run–on slabs.

## 2 Geology of site

The geomorphic history of the site is derived from glaciation, changes in sea level and estuarine activity. The superficial deposits may be assigned to the following broad categories depending on their mode of origin:

Post–glacial deposits (carse clays)
Late–glacial deposits (laminated silty clays)
Glacial deposits

The post–glacial deposits are considered to be of estuarine origin and have limiting elevations above ordnance datum of about 14 and 42 metres respectively.

**Fig 1** shows the location of an old shoreline which forms a distinct physical land feature. The abrupt change in ground level defines the existence of deposits of late–glacial age and the post–glacial carse.

The carse consists of relatively featureless very silty clays which are

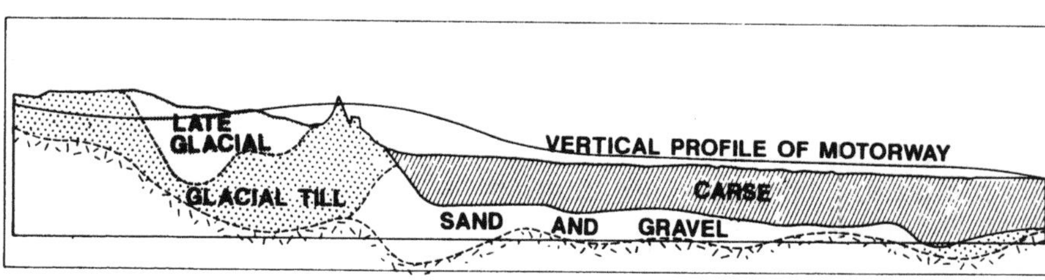


Fig. 1. Location of old shoreline.

analogous to the tidal mud–flats currently exposed at low tide at the Forth River estuary. The carse is considered to be lightly over–consolidated except in the surface layers which generally have a relatively firmer consistency due to desiccation.

An organic deposit known as sub–carse peat defines an important emergence of the land between the late–glacial and the post–glacial submergence.

A complex assemblage of marine deposits underlie the sub–carse peat and consists essentially of various gradations of silts and sands.

The late–glacial deposits which overly the glacial till consist of laminated silty clays, and are typified by alternating bands of silty clay and silt and fine sand. They are generally found within depressions in the irregular and undulating surface of the glacial till and are over–consolidated.

## 3 Ground conditions

An important requirement in the design of earthworks for motorways is an adequate definition of the stress histories and undrained strengths of the superficial deposits, particularly where high embankments will be constructed over soft cohesive soils. The determination of these properties permits a rapid assessment of the probability of instability of embankments and focuses attention on the portions of land which require special consideration and comprehensive investigatory work. The construction measures for avoidance of embankment failure can cause significant delays to the construction programme and all locations of potential instability must be defined at an early stage in the design of the earthworks.

The undrained shear strengths of the post–glacial and late–glacial deposits were obtained from undrained triaxial compression tests and lower bound strengths were deduced from cone penetration tests (CPT), as shown in **Fig 2.**

Where high embankments are to be constructed on soft clay soils, knowledge is required of the in–situ soil permeabilities to permit calculation of the rates of dissipation of the excess porewater pressures generated by the embankment loading. This important information for the prediction of the variations in the factor of safety against embankment failure with time and on the rates of settlement of the embankments was obtained from piezometer measurements.

A critical location for embankment stability on the carse was selected from a study of the site investigation information and three rows of piezometers were installed at positions corresponding to the centre line, mid slope and outwith the toe of the proposed embankment as shown in **Fig 3.**

Typical values of the horizontal and vertical permeabilities of the clay soils are given in **Fig 4.**

The strength parameters of the soils in terms of effective stresses were determined from consolidated undrained triaxial compression tests with porewater pressure measurements and this information was used to predict the factors of safety against embankment failure.

The compressibilities of the soft clays were determined from laboratory consolidation tests on specimens prepared from samples obtained from stationery thin walled piston samplers. Longterm laboratory consolidation tests were also carried out to establish the relationships required for the determination of the secondary consolidation displacements.

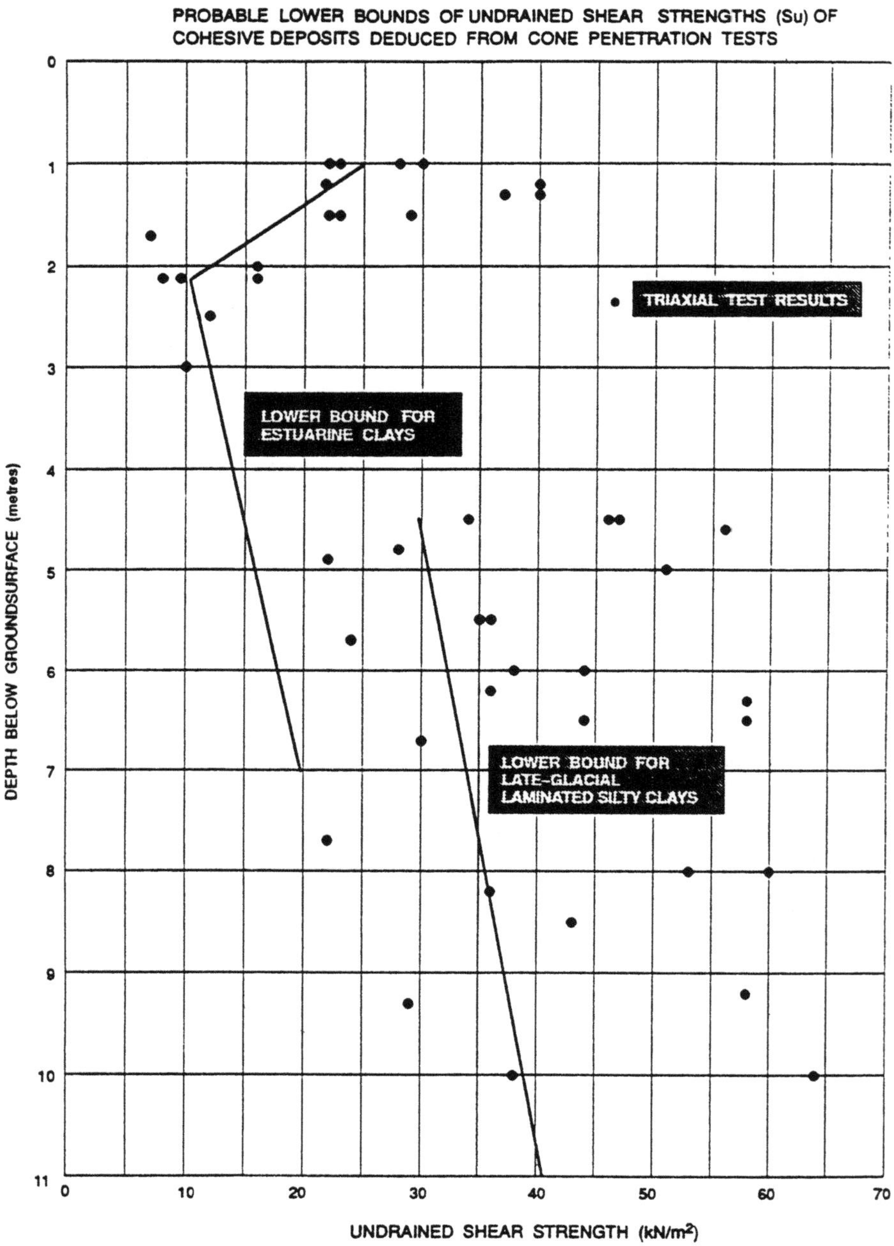


Fig. 2. Undrained shear strengths, deduced from cone penetration tests.

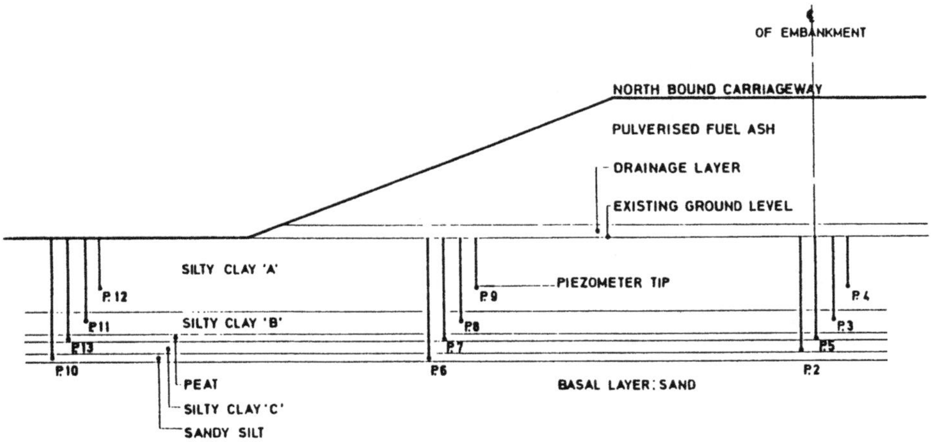


Fig. 3. Location of piezometers.

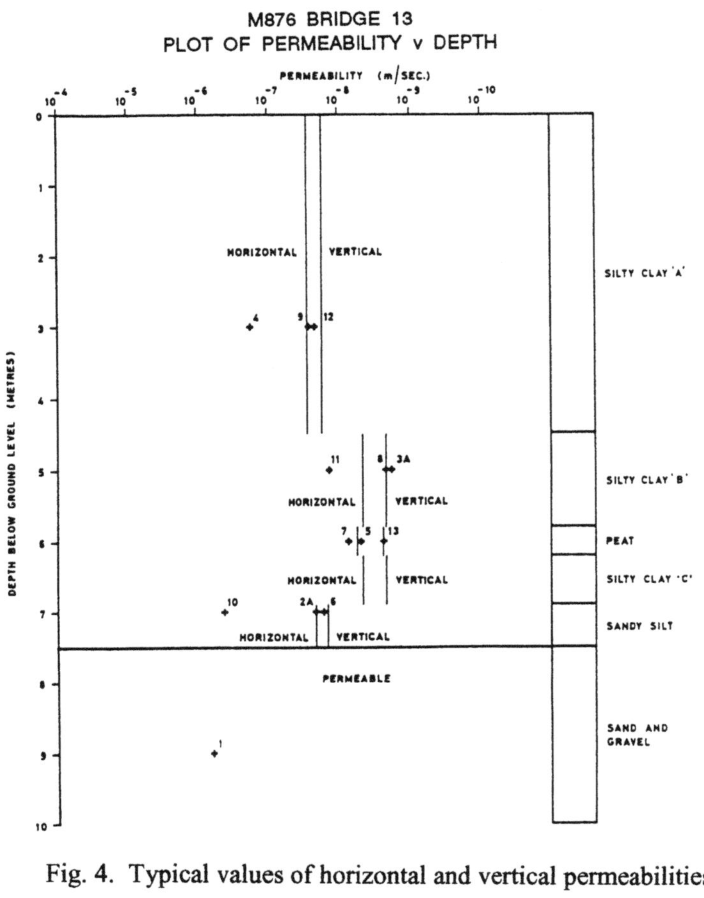


Fig. 4. Typical values of horizontal and vertical permeabilities.

## 4 Predictions of rate of settlement and stability of embankments

The Transport and Road Research Laboratory published a report, Murray (1974) which described a method of analysing the magnitude and rate of settlement on multi–layers soils resulting from a process of two–dimensional consolidation together with porewater pressure generation and dissipation. In addition to the facility for analysing stratified soils the method could take account of the variation of soil parameters during consolidation together with non–linear relationships.

The following integrants of settlement require consideration in the assessment of long–term performance of embankments on soft clay soils.

a. Elastic displacements under embankment loading
b. Primary consolidation displacements due to increases in major principal stresses with the associated generation and dissipation of excess porewater pressures.
c. Shear displacements caused by high ratios of applied stress to limiting stress.
d. Secondary consolidation displacements which commence during the primary phase and continue after dissipation of all excess porewater pressures.

Sufficient laboratory and field test data were accumulated to permit the displacements described in a, b and d to be predicted together with the probable porewater pressures induced in the soils by the embankment loading.

As an important aid to judgement meticulous examination was made of continuous air–dried split samples and the components of the physical structures of the clay soils were carefully identified and recorded. Subsequent examination of colour photographs of the continuous samples taken from the first borehole of each row of piezometer boreholes together with the records of the nature and thicknesses of the physical features of the samples permitted an assessment of the parameters to be used in the analyses.

The physical model adopted for the analytical work is shown in **Fig 5** together with the selected soil characteristics.

The predicted magnitude and rate of settlement under the centreline of the 9 metres high pulverised fuel ash (PFA) embankment founded on the soft carse is

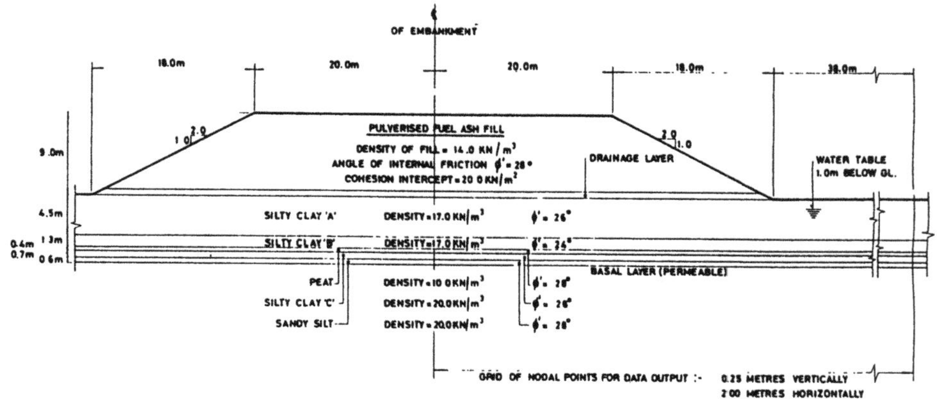


Fig. 5. Physical model and soil characteristics for analysis.

shown on Fig 6, with and without recognition of the influence of secondary consolidation which was found to be only 6% of the maximum total settlement. It was assumed for the purposes of the predictive analyses that the contractor would wish to place the fill material expeditiously and an average rate of construction of 1.12 metres per month was envisaged. In reality the rate of deposition of the PFA was variable and averaged only about 0.68 metres per month requiring a 1.10 year period of construction instead of 0.67 year as permitted by the contract documents.

The allowable rate of construction of the embankments was based on the porewater pressure predictions calculated by the TRRL method and the minimum factor of safety against slope instability at any stage of construction in terms of effective stresses, assessed from the Bishop (1955) method of analysis, was 1.50. The ratio of applied shear stress assessed from Jurgenson (1934) to the limiting stress given by the undrained strength of the carse locally approached unity without any evidence of large shear strains. Some evidence of shear displacement at such high stress ratios would have been expected unless the stressed zones beneath the embankment generally had adequate reserves of strength and were capable of accommodating a local deep zone of plasticity.

## 5 Performances of embankments

The predicted and measured rates of settlement of the 9 metres high embankments founded on the carse are shown in **Fig 7.** The vertical displacements were measured both by precise level and settlement gauges comprising horizontal tubes beneath the embankment. The original settlement predictions are given together with the predictions amended to conform with the slower rate of construction achieved during the contract.

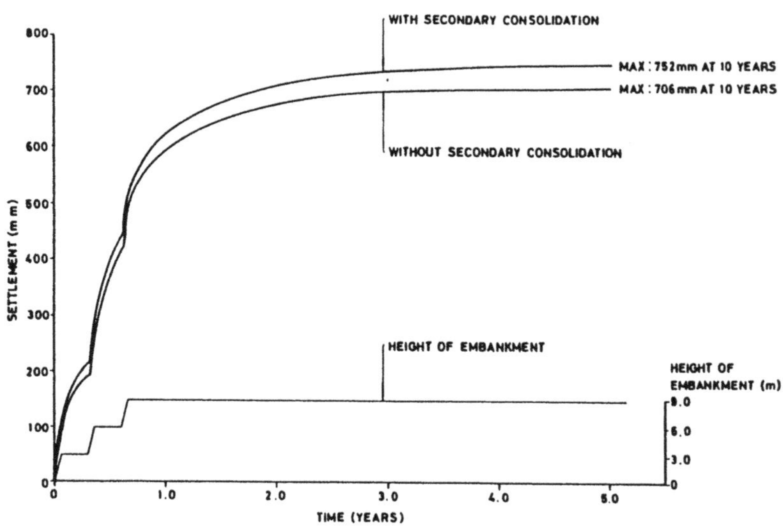


Fig. 6. Predicted settlement under centreline of embankment.

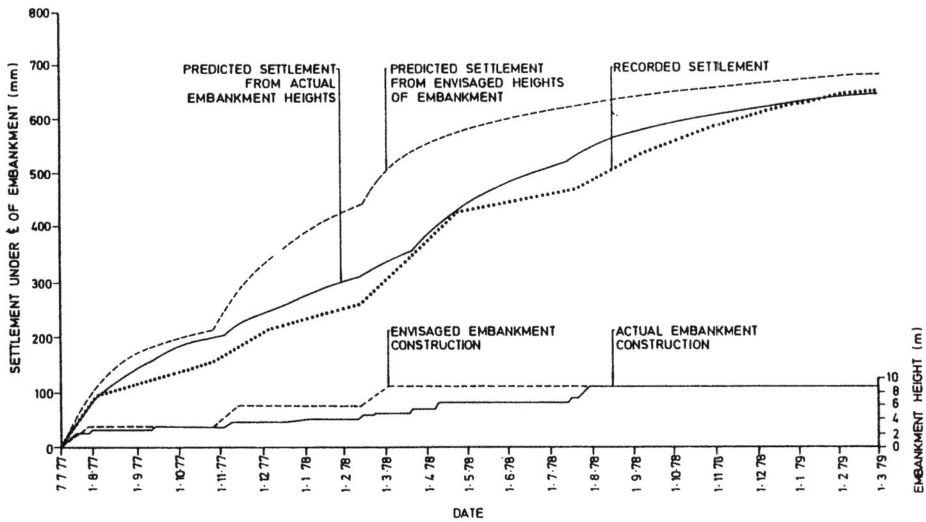


Fig. 7. Predicted and measured settlement

**Fig 8** shows a typical record of measurement of the excess porewater pressures generated relative to the actual rate of deposition of the fill material over a period slightly in excess of one year. This record for a piezometer which was installed at a depth of 3 metres under the centreline of the embankment, shows the rapid increases in porewater pressures on application of embankment loading and the gradual dissipation with time.

The piezometer measurements indicate that only small excess porewater pressures were generated by the construction of the initial 2.50 metres thick layer, and were equivalent to about 0.30 metres of water. The loading intensity imposed on the formation at the centreline of the embankment by 2.50 metres of pulverised fuel ash is approximately 35 kN/m$^2$. The recorded maximum excess porewater pressure of 3 kN/m$^2$ is a relatively small proportion of the applied stress and this behaviour has been observed where the excess porewater pressures were found to be small until the embankment loading exceeded the general over–consolidation stress on the soft carse of about 35 kN/m$^2$.

## 6 Performances of bridges

The previous sections having defined the probable performances of typical embankments on soft clay soils set the background against which we can examine the performances of bridge structures.

We learn much from the performances of earthwork embankments in terms of deformations and it may be of value to bridge designers to consider the criteria listed below which were derived from the field measurements described in this paper.

1. The total settlement of 9 metres high embankments on soft clay soils may be expected to be at least 700mm.

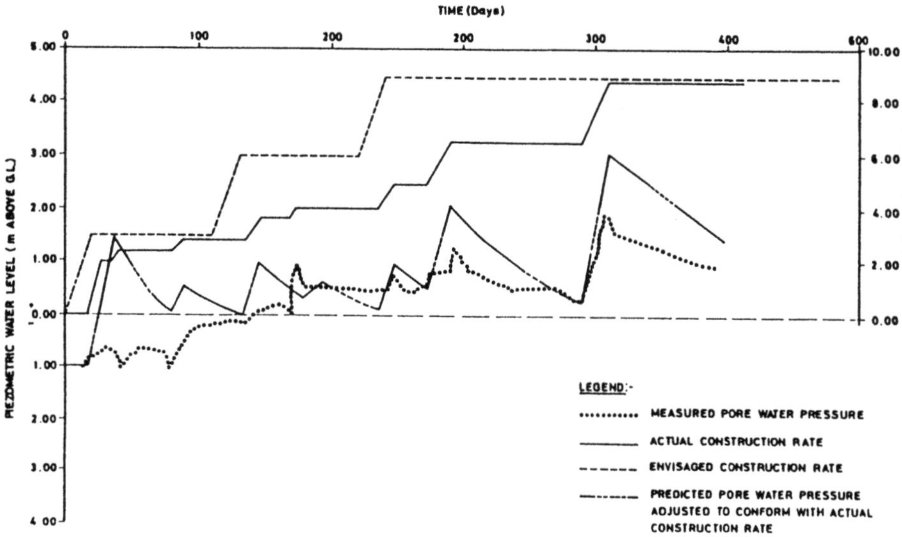


Fig. 8. Typical excess porewater pressures.

2. The time taken to achieve the final settlement of 9 metres high embankments on soft clay soils may be expected to be at least 2 years after completion of the embankment.
3. Lateral displacements of soft clay soils under 9 metres high embankments may be expected to be at least 70 to 150mm.

In order to gain an indication of the situation immediately behind a bridge abutment where an embankment has been formed on soft clays, let us consider the semi–infinite half–plane of homogeneous clay loaded asymmetrically. Let us assume that surfaces of rupture develop in the soft clay soil and the mass rotates clockwise on a cylindrical slip surface which is of unit length normal to the section shown in **Fig 9.** All normal reactions on the cylindrical slip surface pass through the centre of rotation and all tangential reactions are of magnitude (Su) per unit length of arc.

The mathematical solution for the worst case is

$$(q-p) = 5.53Su$$

Where Su is the undrained shear strength of the clay. (Schofield and Wroth, 1968). With reference to **Fig 2,** if we assume that the mean undrained shear strength of the soft clays are 20 kN/m$^2$ an approximation to the maximum height of a PFA embankment before failure occurs is

$$\frac{5.53 \times 20}{14} = 8 \text{ metres}$$

Embankments are constructed, therefore, with 2 to 1 longitudinal side slopes in order to ensure factors of safety ranging from about 1.3 to 1.5 and the soil displacements which are experienced are shown in **Fig 10.**

If embankment fill material is deposited against a bridge abutment, as shown in **Fig 11(a),** the soft clay soils upon which the high embankment is constructed will approach limit stress and plastic behaviour of the soft clays

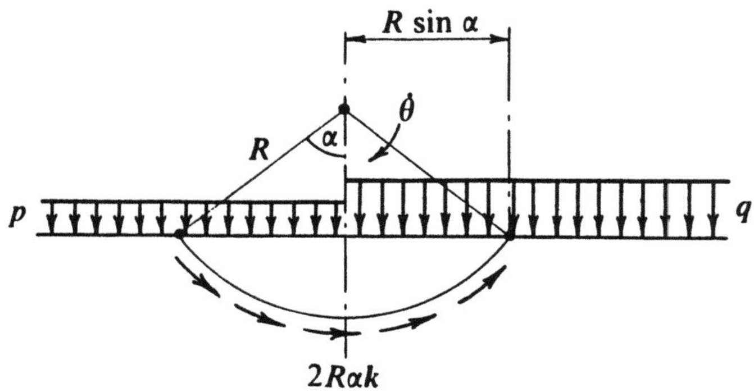


(SCHOFIELD AND WROTH, 1968)

Fig. 9. Assumed model of slip surface.

will be experienced. The lateral displacements of the soft clays in a plastic state impose lateral loads on the piles supporting the bridge abutments.

The lateral loads are considerable and displacements of piled abutments are experienced, as shown in **Figs 12** and **13.**

Advanced earthworks which involve construction of embankments some one to two years before construction of bridges may improve embankment performance, but

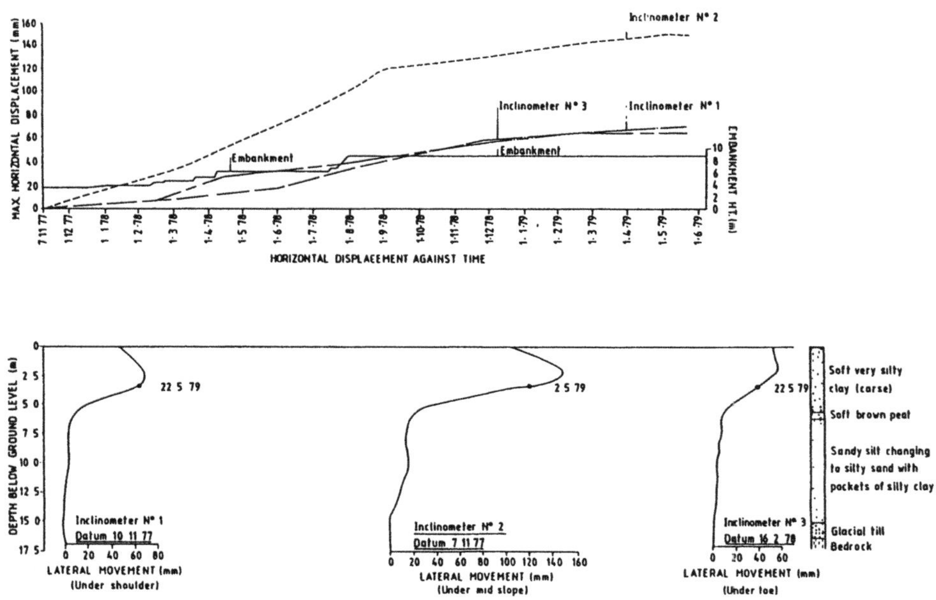


Fig. 10. Inclinometer curves for M876.

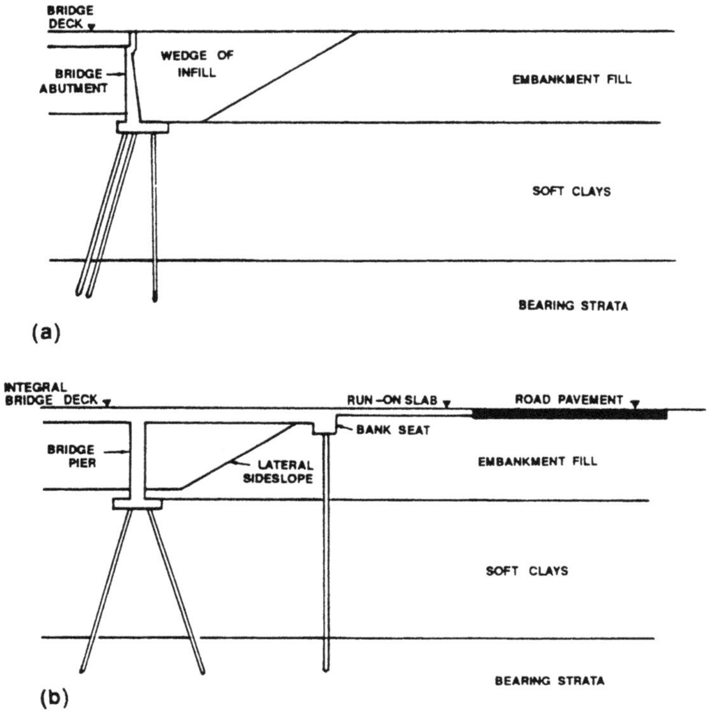


Fig. 11. Alternative methods of constructing embankments.

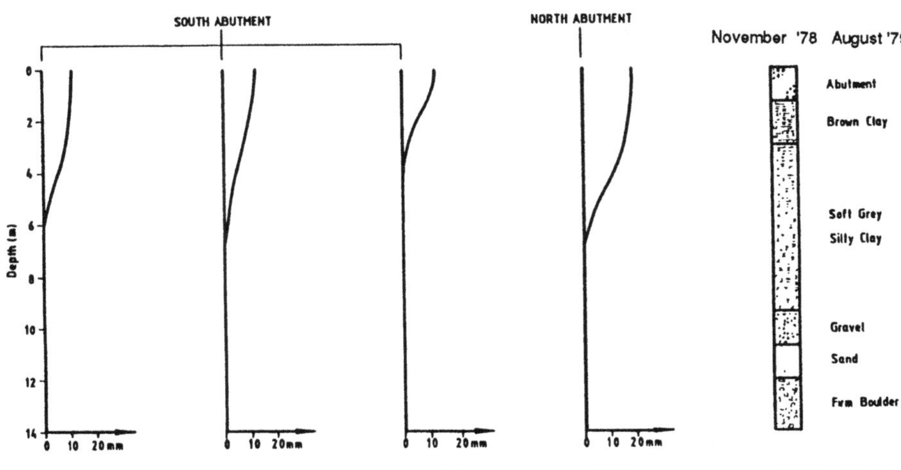


Fig. 12. Displacement of piled abutments of M876 Bridge 20, measured with inclinometers.

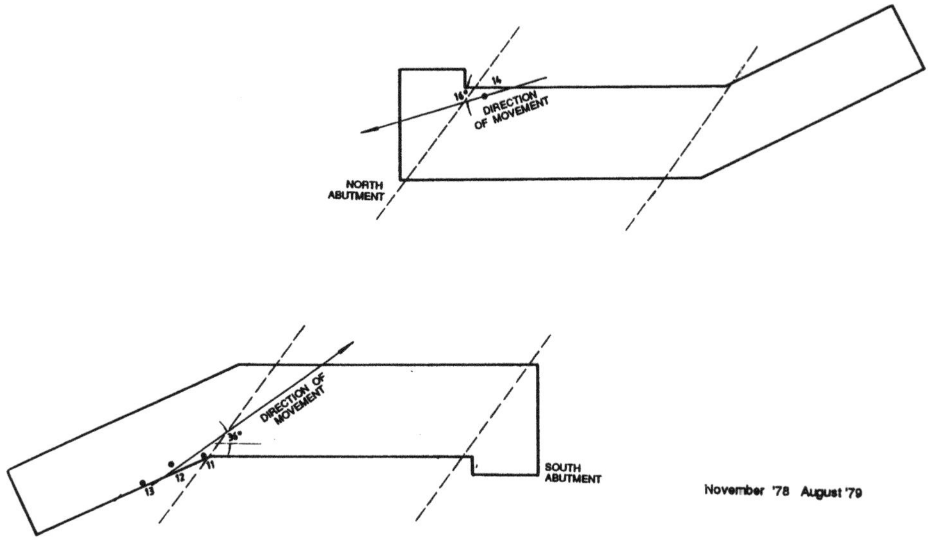


Fig. 13. Direction of abutment movement, M876 Bridge 20, as indicated by inclinometers.

do not relieve the near–limit stresses induced by the final infilling of the wedges behind the bridge abutments.

The advanced earthworks could be extended over individual bridge sites to pre–load the soft clays and reduce the effect of infilling of the wedges behind the abutments. There would, however, be a cost and time penalty in adopting this solution, including the problem of early land acquisition.

The alternative solution of constructing additional bridge decks supported on bank seats and the construction of lateral side slopes to minimise soil displacements, as shown on Fig **11(b)**, may be adopted together with the use of 'soft' piles to support bank seats.

The concept of run–on slabs was developed as an adjunct to integral bridges.

Integral bridges may be described as those constructed without movement joints over the entire length of the bridge deck. In order to create a transition between the continuous concrete bridge deck and the road pavements at either end of the bridge, run–on slabs may be provided, as shown on **Fig 11(b)**.

Joints are required between run–on slabs and reinforced concrete road pavements in order to accommodate movements, but the interfaces between bituminous road pavements and run–on slabs are not provided with special joints.

Degradation of the road pavement at these interfaces is inevitable, but maintenance costs apparently are considered acceptable.

In situations where embankments have been formed on soft clays, **Fig11(a)**, run–on slabs cannot be used over the lengths of wedges of infill behind bridge abutments since they do not eliminate the problem of soil displacements at depth.

In order to overcome the problem of soil displacements behind bridge abutments the author proposed in 1976 the solution of using geotextiles and individual piles to support the wedges of fill.

It was proposed that the portions of embankments immediately behind bridge abutments be supported on individual piles arranged such that the percentage of total embankment load carried by the piles would reduce with increasing distance from the abutment. The bridges would be designed with normal abutments without additional side spans and the need for embankment pre-loading would be removed. A significant benefit was achieved by the removal of any restrictions on rate of infilling behind bridge abutments.

A comprehensive description of the support system was presented by Reid and Buchanan (1984).

The Author does not recommend the use of run-on slabs in situations where infill would be placed on soft clays immediately behind bridge abutments and where large soil displacements would be expected. The research being conducted by the University of Cambridge should be supported by Highway Authorities.

## 7 References


Bishop, A.W. (1955). The use of the Slip Circle in the stability analysis of slopes. **Geotechnique** 5 (7), 17.

Francis, E.H., Forsyth, I.H., Read, W.A. and Armstrong, M. (1970) – **Geology of Stirling District.**

Hambly E.C. (1991) **Bridge Deck Behaviour.** E&FN Spon.

Hvorslev, M.J. (1951) Time lag and soil permeability in groundwater observations. **Bulletin No 36, Waterways Experiment Station, Corps. of Engineering,** Vicksburg, Mississippi.

Jurgenson, L. (1934). The application of theories of elasticity and plasticity to foundation problems. **Journal of the Boston Society of Civil Engineers.**

Murray, R.T. (1973) Embankments constructed on soft foundations: Settlement studies near Oxford. **TRRL Report No 538.**

Murray, R.T. (1974) Two dimensional analysis of settlement by computer programme. **TRRL Report No 6117.**

Reid W.M. and Buchanan N.W. (1984) Bridge approach support piling. **Proceedings of conference on advances in piling and ground treatment.** The Institution of Civil Engineers, March 1983.

Schofield, A. and Wroth, P. **Critical state soil mechanics.** McGraw–Hill.

Sissons, J.B. (1971) Geomorphology and foundation conditions around Grangemouth. **Quarterly Journal, Engineering Geology** 3, 183–191.

Thorburn, S. (1976) The static penetration test and the ultimate resistance of driven piles in fine–grained non–cohesive soils. **The Structural Engineer** 54, 6.

Thorburn, S. (1988) Soil Interation, The Real Behaviour of Structures. **Institutions of Structural and Civil Engineers.**

# REPORT ON SESSION EIGHT

D.W. QUINION

Dr Springman described the inferences from the recent centrifuge testing examination of the soils behaviour behind the end abutments or retaining walls of integral bridges. The next stages of the testing would examine changing the important parameters. The effects of creep were suggested as likely to ameliorate the extent to which bending moments would increase with increasing abutment stiffness.

Mr Low emphasized the considerable benefits provided by the horizontal stiffness of the highway pavement which could double the stiffness provided by the abutment. It would be vulnerable to any trenching carried out across the pavement adjacent to an abutment. He suggested a diagrammatic relationship which demonstrated the interaction between the force in the deck and the displacement at each abutment (Fig. 4 in the paper).

Dr Thorburn described the problems behind the abutment when using soft clay fills. Embankments are usually constructed in stages with or without provision to speed the dissipation of pore pressures. Bridges are comparatively rigid structures. The problems occur from the need to backfill quickly between a bridge and an embankment which usually has different settlement or consolidation characteristics. It is important to carry a run-on slab well back from the abutment. There seemed to be agreement that replacing the abutment with a pier and carrying the deck to a bank seat from which a run-on slab would act would overcome such problems. Alternatively the wedge of filling behind an abutment would be stiffened by the inclusion of soft piles under a run-on slab.

Bridge decks and pavements should ideally have similar temperature-related characteristics. Mr Low emphasized the advantages in this respect from the use of lightweight concrete for integral bridges.

# CONCLUDING SUMMARY

A.M. LOW

The Chairman started the discussion by offering some views of a contractor, a viewpoint which had not been represented in the presentations:

- Bridges must be maintainable and the requirements for maintenance should be included in the basic concept.
- Bridges must be safe to build and formal quality management systems for the design had an important role to play in achieving this safety.
- Any critical requirements for construction should be clearly stated from the start in the documentation of the design.

To open up the discussion he asked if there were any deserving topics which had been omitted thus far. It was suggested that slender columns might not have been covered. A truly joint-free bridge would also have no joints between the deck and the columns which would be monolithic. This is a common design feature which results in columns which are slender to minimise the stresses due to temperature changes. The design of such columns requires careful thought because of the contradictory requirements for vertical strength and sway flexibility. Current rules can be unduly conservative. What is needed is clear guidance from someone who fully understands the background to the various code rules for slender columns.

Another topic which had not been covered was fatigue. Deck continuity introduces reversals into the moments as a vehicle passes and this can greatly increase the fatigue damage.

The Chairman also asked for topics which required further work. The inter-relationship between crack control, waterproofing and durability was identified as a topic needing further study, particularly for the slabs of composite decks which had been made continuous. In the hogging regions they were in axial tension over their full depth.

A request was made for more design studies of joint-free bridges to be published to encourage designers. It was reported that the Steel Construction Institute would be looking into this.

There was also a request for more money to be available for instrumentation, both on full-scale bridges and on the centrifuge models in Cambridge which had been described in the previous session. This led to further discussion of the centrifuge experiments. One participant felt that it should model fills which were more compressible. Another took the opposite view that it should model the stiffer response of cemented fills.

## Final words

The Chairman's final words were "thank you". He thanked all those whose efforts had made the colloquium happen.

Brian Pritchard then closed the proceedings. He was pleased that the Henderson Colloquium had maintained its international involvement. The participants had come from eight countries. He considered that the presentation of the papers had been a success, covering very many aspects of the topic. He also felt that the discussions had been successful in building on the contributions of the papers. He announced that, following the usual practice for Henderson Colloquia, an open meeting would be held at the Institution of Structural Engineers in London to report the proceedings and to allow those not present to add their contributions.

He ended with some reflections on how the continuing development of technology would affect the design of bridges. He had recently visited Germany where he had been introduced to the "intelligent bridge" which monitored its own strains and corrosion and he wondered if this would become standard practice.

# AUTHOR INDEX

# SUBJECT INDEX

For Product Safety Concerns and Information please contact our EU representative GPSR@taylorandfrancis.com
Taylor & Francis Verlag GmbH, Kaufingerstraße 24, 80331 München, Germany

www.ingramcontent.com/pod-product-compliance
Ingram Content Group UK Ltd.
Pitfield, Milton Keynes, MK11 3LW, UK
UKHW022003180926
475701UK00003B/174

* 9 7 8 0 3 6 7 8 6 4 0 6 4 *